Handbook of Games
and Simulation Exercises

Handbook of Games
and Simulation Exercises

Edited by

G. I. GIBBS

E. & F. N. SPON LTD

LONDON

First published in 1974 by
E. & F. N. Spon Limited
11 New Fetter Lane, London EC4P 4EE

© *1974 G. I. Gibbs*

Printed in Great Britain by William Clowes & Sons Limited
London, Colchester and Beccles

ISBN 0 419 10780 0

To Jean

'In the modern marriage, both partners
choose each other freely as persons.'

R. Fletcher

For which I thank providence

Preface

No book of this nature is entirely the work of the author. The writing of this book was made possible by the unstinting help of fellow gamers—to them my thanks. In particular I would like to express my deep sense of gratitude to the Council of the Society for Academic Gaming and Simulation in Education and Training, to Phillip Read who oiled the mechanism of publication, and especially to Dr David Teather for his constant encouragement.

G. IAN GIBBS

Moreton-in-Marsh

Contents

Preface *page* vii

1. Introduction 1

2. Analysis 4

  2.1 The vocabulary of gaming and simulation 4
  2.1.1 Communication 4
  2.1.2 Glossary 4
  2.1.3 Bibliography 9
  2.2 Gaming and simulation literature 11
  2.2.1 Bibliographies 11
  2.2.2 Magazines and periodicals 13
  2.2.3 Books 17

3. Synthesis 35

  3.1 The making of a game 35
  3.1.1 Aims 35
  3.1.2 Context 37
  3.1.3 Target population 37
  3.1.4 Models 37
  3.1.5 Scenario 38
  3.1.6 Information flow 38
  3.1.7 Cast list 39
  3.1.8 Role elaboration 39
  3.1.9 Player objectives 39
  3.1.10 Statement of rules 39
  3.1.11 Description of constraints 39
  3.1.12 Materials and equipment 40
  3.1.13 Evaluation 40
  3.1.14 Reprography 40
  3.1.15 Bibliography 40
  3.2 Training and sustaining 43
  3.2.1 Course organizers 43
  3.2.2 Associations, clubs and societies 45
  3.3 Using games and simulations 48

x : Contents

3.3.1	*Teaching*	49
3.3.2	*Evaluation*	49
3.3.3	*Research*	49
3.3.4	*Bibliography*	50

4. Prognosis — 52

4.1	Problems and solutions	52
4.2	Bibliography	53

5. Register of Games and Simulations — 54

5.1	Games and simulations	54
5.2	Subject index	185
5.3	Author index	190
5.4	Manufacturers and suppliers	197
5.5	Appendix	221

1 Introduction

Currently information about games and simulations appears in small, specialized amounts in a wide variety of sources. A much lamented situation exists in which there is no *one* British source of information. This book is an attempt to fill that gap. The fact that the information is neither complete nor comprehensive is patent. However, on the principle 'nothing ventured, nothing gained' it has been thought worthwhile publishing the information at hand. I apologize for any inaccuracies, omissions, etc. Coupled with the apology I wish to appeal to all those who have better quality and/or more information to let me know so that, hopefully, future editions may be improved.

The information included is intended to help the educator (trainer) to involve himself, and subsequently his students, in gaming and simulation. A guide to the jargon used has been included so that communication can be based on a common language. Sources of information about gaming have been quoted, both published and organizational. Details of training available are given, as are guides to making and using games and simulations. Problems and some possible solutions are discussed. Finally, not least in importance are the data about games and simulations which are available.

It is fairly easy to become over-concerned about which games and simulations should be included as academic, and which left out. I have operated on the simple principle that if there is a possible educational benefit to be derived from a game it should be included.

Information about specific games and simulations has been given in the following form:

Title; Year of publication; Country of origin
Author; Publisher
Target population
Number of players. Approximate duration
Major aims/objectives/content

For example:

PURCHASING; 1968; USA
E. Rausch; Didactic Systems Inc (also Science Research Associates)
Experienced buyers, management
3-5; 2 hours
Insights into the major functions of purchasing

COMPARISON OF U.K. AND U.S.A. SCHOOL SYSTEMS

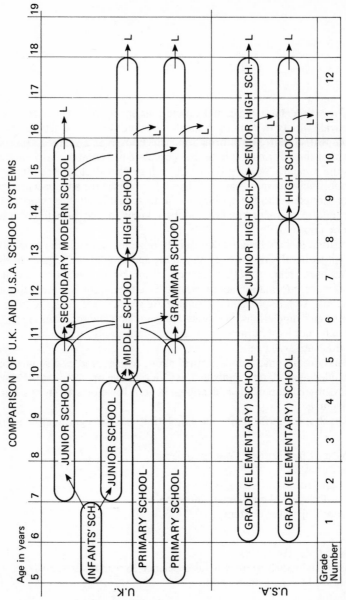

Notes:

1. In the U.K. groups of children up to the age of ten or eleven are usually called classes, after eleven they are called forms.
2. In the U.S.A. groups of children are called grades.
3. The correlation between grade number and age is subject to local and individual variations but serves as a rough yardstick.
4. L means leaves school, either for employment or further education.

A number of the school-level games included designate their 'target population' by terms which have different meanings in America and the U.K. A brief explanation of the two educational systems may therefore be useful.

The American system consists of twelve years of full-time education. The way this period is divided varies from State to State. A common pattern is for the first eight years to be spent in Grade (Elementary) School followed by a further four years in High School. Even more common is a pattern of six years in Grade School, three years in Junior High School and three years in Senior High School.

The pattern of American educational levels is compared with the British system in the diagram on the opposite page.

For a number of games the data given are incomplete, but they have been included in the hope that further information will be forthcoming, either from the authors or publishers or from educators who prove more successful than I at tracing references.

2 Analysis

2.1 The vocabulary of gaming and simulation

'True wit is nature to advantage dressed,
What oft was thought, but ne'er so well expressed.'
Alexander Pope

'... words strain,
Crack and sometimes break, under the burden,
Under the tension; slip, slide, perish,
Decay with imprecision ...'
T. S. Eliot

2.1.1 Communication

The development of gaming techniques has taken place in a number of widely disparate fields, e.g. town planning, politico-military strategic evaluation, mathematics. Generally these fields are far enough removed from one another for little contact to be made between them. Consequently, each field has developed its own terminology as the need has arisen. As gamers from the various fields feel the need to communicate, problems of interpretation are occurring with increasing frequency due to the overlapping but dissimilar uses of jargon terms in the various fields.

This glossary has been compiled on the basis of what appears, from the literature, to be the most widely accepted interpretation of each term. The source of each definition is given, together with the sources of alternative definitions where these seem appropriate. The numbers in bold after the definition refer to the main source, other numbers refer to additional sources.

2.1.2 Glossary

ALGORITHM   'By algorithm we mean the usual automatic procedure of solving a problem ...' (**1**, 2)

ANALOGUE MODEL   '... analogue models are those in which one property is represented by another ...' (**3**, 4)

BUSINESS GAME	'A dynamic teaching device which uses the sequential nature of decisions as an inherent feature of its construction and operation.' **(5, 6, 7)**
CASE STUDIES	'. . . present a problem for discussion, derive the problem from previous events and typically deal with problems involving other people.' **(6, 2, 3, 7)**
COMMUNICATION THEORY	'. . . the direction of flow of information (who communicates to whom) and . . . the structure of the communication network and the content of the messages carried . . . ' **(8)**
COMPANY GAMES	'Where most of the functions of a company are simulated in a game, but the participants are only concerned with the internal operations and consequences of decisions.' **(5)**
COMPETITION	'Striving with another for an object or vying with another in a quality.' **(9)**
COOPERATION	'Working together to produce an effect.' **(9)**
CRISIS GAME	'. . . a device for crudely simulating the decision making process . . . ' **(10)**
CYBERNETICS	'. . . the science of proper control within any assembly that is treated as an organic whole.' **(11, 2, 12)**
DECISION RULES	'Criteria on the basis of which decisions are made about resources and information to be transmitted or received or what actions are to be taken.' **(13)**
DETERMINISTIC SYSTEM	'. . . can be determined in advance where every inter-relationship between every part can be prescribed . . . ' **(14)**
EDUCATIONAL TECHNOLOGY	'. . . the technique of instigating problems of learning, organization, control and logistics by logical analysis and programme construction.' **(2, 15, 40)**

EVALUATION	'The act of determining the value, or of ascertaining some means of stating the value, of an unknown in terms of that which is known.' (**16**, 2, 42)
FEEDBACK	'The transmission of information from the end result of a process to a centre controlling that process . . . ' (**17**, 2, 18, 19)
FREE KRIEGSPIEL	'War games in which a referee evaluates decisions made.' (**5**)
FUNCTIONAL GAMES	'Business games covering only one (or possibly two or three) functions performed within the simulated company.' (**5**)
GAME	'An activity carried out by cooperating or competing decision-makers seeking to achieve, within the rules, their objectives.' (**9**, 20, 21)
GAME THEORY	'Theoretical models in which the player has some control over outcomes but not complete control because other players and chance will influence events.' (**22**, 30)
HOMEOSTASIS	'The regulation of variables important to the survival or well being of an organism.' (**12**)
HEURISTIC PROCESS	'. . . by a quasi-algorithm we mean the intelligent guess or "short cut" . . . ' (**1**)
ICONIC MODELS	'. . . those which represent the real world with only a change in scale.' (**3**, 4)
IN-BASKET	*See* IN-TRAY
INCIDENT PROCESS	'. . . a variant on the case-study, in which the participants seek for additional information as they work . . . ' (**3**)
INFORMATION THEORY	'A system of measurement used originally by communication engineers for quantifying the channel capacity of a system.' (**17**)
INTERACTION	'. . . flow of resources and information to and from each player.' (**13**)
IN-TRAY	'. . . a technique in which one person is asked to consider and act on a set of issues . . . ' (**3**, 7)

LOGISTICS	'. . . the supply and deployment of resources for a particular purpose.' **(2)**
MANAGEMENT GAMES	*See* BUSINESS GAMES
MICRO-TEACHING	'. . . a method of giving practice and feedback in teaching by use of a videotape playback.' **(23**, 24)
MINIMAX	'. . . when the pay-off for A equals that of B.' **(22**, 30)
MODEL	'A model is a method of transferring some relationship or process from its actual setting to a setting where it is more conveniently studied.' **(25**, 3, 26, 27, 28, 39)
MONTE CARLO METHOD	'. . . a situation in which a difficult but determinate problem is solved by resort to a chance process.' **(29)**
NEW KRIEGSPIEL	'War games, which are either free or rigid, using real maps.' **(5)**
NON-ZERO-SUM GAMES	'. . . in which the winner's gain is not necessarily at the cost of the loser.' **(20)**
OBJECTIVES	'. . . a statement of the aim, purpose or goal of some activity.' **(2**, 41)
OPERATIONAL RESEARCH	'The scientific approach to complex problems arising in the direction and management of systems of men, machines, materials and money in industry, business, government and defence.' **(27)**
PARETO OPTIMUM	'The best strategy in such games (non-zero-sum) is one that maximizes the total wins of all players.' **(20)**
PAY-OFF	'. . . the benefits accruing from winning a game.' **(9**, 29)
PROBABILISTIC SYSTEM	'. . . (In which one) cannot be certain on how the total system will behave . . . ' **(14)**
RANDOM NUMBERS	'. . . the selection of these numbers must be left to chance.' **(30)**
RIGID KRIEGSPIEL	'War game with rigid rules and dice are used to introduce chance.' **(5**, 26)

ROLE PLAY

'. . . role-playing is simply taking or trying out first person action in a learning situation.' (**7**, 6, 16)

SAMPLE STUDY

See CASE STUDY

SCENARIO

'All the necessary background information for a game or simulation.' (**31**, 13)

SHOWDOWN GAME

'. . . simultaneous player actions depending little on uncertainties of opponents' moves.' (**20**)

SIMULATION

'Dynamic representation which employs substitute elements to replace real or hypothetical components.' (**32**, 9, 21, 31, 33, 34, 35, 38)

SIMULATORS

'. . . a simulator is designed to represent a real situation, to provide a student with control over that situation, and to vary conditions during training, so that the task can be made progressively more difficult.' (**6**, 2, 38)

SOCIODRAMA

'Role-playing as a means of seeking a solution to a social problem.' (**16**)

STOCHASTIC GAMES

'The games are genuinely stochastic if the outcome is determined by the use of random numbers.' (**5**)

STRATEGIC GAME

'. . . sequential player actions that are mutually responsive and stress need for predicting opponents' moves.' (**20**)

STRATEGY

'A strategy is a plan for arriving at a pre-determined goal at minimum cost.' (**36**)

SUB-SYSTEM

'An identifiable part of a total system that performs a major function in achieving the stated mission objectives for the total system.' (**37**, 2)

SYMBOLIC MODELS

'. . . symbolic models are those in which symbols represent objects or properties.' (**3**, 4)

SYSTEM
'A set of units with relationships among them, an intercommunicating network of attributes or entities forming a "complex whole", "an assembly of attributes", not a thing but a list of variables.' **(18**, 2)

SYSTEMS THEORY
'Conceptual construct of behavioural systems subject to control by human beings, which is generalizable.' **(9)**

TACTICS
'Art of disposing manpower in actual contact with opponents in pursuance of an end in a competitive situation.' **(9)**

TARGET POPULATION
'The student population at which learning materials are aimed.' **(9)**

TEWTS
'Tactical exercises without troops.' **(5)**

TOTAL ENTERPRISE GAMES
See BUSINESS GAMES

WAR GAMES
'. . . a crisis game set in a military (naval) environment.' **(9)**

WIN CRITERIA
'Scoring rules based on the degree to which players achieve their objectives with efficient utilization of resources.' **(13)**

ZERO-SUM GAMES
'. . . games of pure competition or conflict (generally called zero-sum games because the sum of one side's gain and the other's loss will always be 0 . . .' **(22**, 20, 30)

2.1.3 Bibliography

1   GEORGE, F. (1970) *Models of Thinking*. London, Allen & Unwin.
2   BLAKE, C. S. (1972) in *APLET Yearbook of Educational and Instructional Technology 1972/73* (Ed. A. J. ROMISZOWSKI). London, Kogan Page.
3   WALFORD, R. (1969) *Games in Geography*. London, Longmans.
4   HODGE, P. (1970) in *Aspects of Educational Technology IV* (Ed. A. C. BAJPAI and J. F. LEEDHAM). London, Pitman.
5   LOVELUCK, C. (1963) *Notes on the Construction, Operation and Evaluation of Management Games*. Weybridge, Management Games Ltd.
6   DAVIES, I. K. (1971) *The Management of Learning*. Maidenhead, McGraw-Hill.
7   ZOLL, A. A. (1966) *Dynamic Management Education*. Reading, Mass., Addison-Wesley.

8 KATZ, D. and KAHN, R. L. (1971) in *Management Information Systems* (Ed. T. W. McCRAE). Harmondsworth, Penguin.

9 GIBBS, G. I. (1972) *Glossary Notes.* Moreton-in-Marsh (mimeo).

10 BLOOMFIELD, L. P. and WHALLEY, B. (1965, winter) *The Political-Military Exercise: A Progress Report.* Orbis.

11 BEER, S. (1960) *Cybernetics and Management.* London, E.U.P.

12 PASK, G. (1961) *An Approach to Cybernetics.* London, Hutchinson.

13 GLAZIER, R. (1969) *How to Design Educational Games.* Cambridge, Mass., Abt Associates Inc.

14 ROMISZOWSKI, A. J. (1970) *The Systems Approach to Education and Training.* London, Kogan Page.

15 REID, R. L. (1969) in *Media and Methods: Instructional Technology in Higher Education* (Ed. D. UNWIN). Maidenhead, McGraw-Hill.

16 GARVEY, D. M. (1971) in *Educational Aspects of Simulation* (Ed. P. J. TANSEY). Maidenhead, McGraw-Hill.

17 BIGGS, J. B. (1968) *Information and Human Learning.* Melbourne, Cassell.

18 RICHMOND, W. K. (1969) *The Education Industry.* London, Methuen.

19 ANNETT, J. (1969) *Feedback and Human Behaviour.* Harmondsworth, Penguin.

20 ABT, C. C. (1970) *Serious Games.* New York, Viking Compass Press.

21 UNWIN, D. (1969) in *Media and Methods: Instructional Technology in Higher Education* (Ed. D. UNWIN). Maidenhead, McGraw-Hill.

22 WILSON, A. (1970) *War Gaming.* Harmondsworth, Penguin.

23 LEITH, G. O. M. in *Aspects of Educational Technology V* (Ed. D. PACKHAM *et al.*) London, Pitman.

24 McALEESE, W. R. and UNWIN, D. (1970) *Microteaching: A Bibliography.* Coleraine, New University of Ulster.

25 DE BONO, E. (1969) *The Mechanism of Mind.* Harmondsworth, Penguin.

26 TANSEY, P. J. and UNWIN, D. (1969) *Simulation and Gaming in Education.* London, Methuen.

27 MINISTRY OF LABOUR (1967) *Glossary of Training Terms.* London, H.M.S.O.

28 ABT, C. C. (1964, August) *War Gaming.* International Science and Technology.

29 SINGH, J. (1968) *Operations Research.* Harmondsworth, Penguin.

30 REICHMAN, W. J. (1964) *Use and Abuse of Statistics.* Harmondsworth, Penguin.

31 BOARDMAN, R. and MITCHELL, C. R. (1971) in *Educational Aspects of Simulation* (Ed. P. J. TANSEY). Maidenhead, McGraw-Hill.

32 DIMITRIOU, B. (1971) in *Feedback on Instructional Simulation Systems* (Ed. R. H. R. ARMSTRONG and J. L. TAYLOR). Cambridge, Cambridge Institute of Education.

33 SYSTEM DEVELOPMENT CORP. (1965, April) *Simulation: Managing the Unmanageable.* Palo Alto, Calif., System Development Corporation Magazine.

34 TWELKER, P. A. (1970) in *The Guide to Simulation Games for Education and*

Training (Ed. D. W. ZUCKERMAN and R. E. HORN). Cambridge, Mass., Information Resources Inc.

35   TWELKER, P. A. (1971) in *Educational Aspects of Simulation* (Ed. P. J. TANSEY). Maidenhead, McGraw-Hill.

36   HUNT, E. B. (1962) *Concept Learning: An Information Processing Problem.* New York, Wiley.

37   CORRIGAN, R. E. and KAUFMAN, R. A. (1965) *Why System Engineering.* Palo Alto, Calif., Fearon.

38   GAGNÉ, R. M. (1965) in *Training Research and Education* (Ed. R. GLASER). New York, Wiley.

39   NEIL, M. W. (1970) in *Aspects of Educational Technology IV* (Ed. A. C. BAJPAI and J. F. LEEDHAM). London, Pitman.

40   HUBBARD, G. (1970) in *Aspects of Educational Technology IV* (Ed. A. C. BAJPAI and J. F. LEEDHAM). London, Pitman.

41   McMULLEN, I. (1970) in *The Teacher as Manager* (Ed. G. TAYLOR). London, N.C.E.T.

42   ERAUT, M. (1970) in *The Teacher as Manager* (Ed. G. TAYLOR). London, N.C.E.T.

2.2   Gaming and simulation literature

'. . . reading should be regarded mainly as the art of getting ideas from script or print, ideas that can please, instruct, elevate or inspire.'

A. F. Watts

'. . . it is true that reading will often awake new interests as well as nourishing existing ones.'

Lady Plowden, *et al.*

2.2.1   Bibliographies

There are a number of bibliographies published in the field, some general and some specific to particular areas of study. A selection of these is given below.

Annotated Bibliography of Simulation in the Social Sciences. Hartman, J. (1966). Iowa State University, Iowa.

A Selected Bibliography of Simulations of International Relations. Fabri, D. A. (1967). Peace Research Centre, Lancaster.

Bibliography of Business Games. Loveluck, C; B.L. Industrial Training Aids Ltd.

Bibliography of Games-Simulations for Teaching Economics and Related Subjects. Joint Council on Economic Education (1968).

Bibliography of Simulations: Social Systems and Education. Werner, R. and Werner, J. (1969). Western Behavioural Sciences Institute.

Bibliography on Simulation, Gaming, Artificial Intelligence and Allied Topics. Shubik, M. (1960). Journal of the American Statistical Association.

Bibliography on the Use of Simulation in Management Analysis. Malcolm, D. G. (1960). Operations Research.

Bibliography on War Gaming. Riley, V. and Young, J. P. (1957). John Hopkins University, Baltimore.

Business Games. British Institute of Management.

Business Games Handbook. Graham, R. G. and Gray, C. F; Didactic Systems Inc.

Business Gaming: A Survey of American Collegiate Schools of Business. Dale, A. G. and Klasson, C. B. (1964). University of Texas.

Computer Simulation. Ministry of Defence (Army).

Directory of Educational Simulations, Learning Games and Didactic Units. Klietsch, R. G. and Wiegman, F. B; Instructional Simulations Inc.

Game Theory. Ministry of Defence (Army).

Glossary of Terms used in Gaming and Simulation. Defence Operational Analysis Establishment.

Instructional Materials. University Council for Educational Administration (1967).

Instructional Simulation Systems: An Annotated Bibliography. Twelker, P. (1969). Waldo Hall 100.

Microteaching: A Bibliography. McAleese, W. R. and Unwin, D. (1970). New University of Ulster, Coleraine.

Model Soldiers. Ministry of Defence (Army).

Operational Gaming and Simulation in Urban Research: An Annotated Bibliography. Duke, R. D. and Schmidt, A. H. (1965). Michigan State University.

Select Bibliography. Unwin, D. (1971). Educational Aspects of Simulation.

Selective Bibliography on Simulation Games as Learning Devices. Boocock, S. S. and Schild, E. O. (1968). Simulation Games in Learning.

Simulation: An Annotated Bibliography. Hogan, A. J. (1968). Social Education.

Simulation and Gaming in Business and Economics in the 1960s. Johnson, E. R. (1969). University of Iowa.

Simulation and Gaming in Education, Training and Business: A Bibliography. Tansey, P. J. and Unwin, D. (1971). Society for Academic Gaming and Simulation in Education and Training.

Simulation, Role-Playing and Sociodrama in the Social Studies: An Annotated Bibliography. Garvey, S. K. (1967). Emporia State Research Studies.

Social Science Instructional Simulation Systems: A Selected Bibliography. Taylor, J. L. University of Sheffield.

Sources in Simulation and Academic Gaming: An Annotated Bibliography. Tansey, P. J. and Unwin, D. (1969). British Journal of Educational Studies.

The Guide to Simulation Games for Education and Training. Zuckerman, D. W. and Horn, R. E. (1970). Information Resources Inc.

The Simulation of Cognitive Processes II: An Annotated Bibliography. Simmons, P. L. and Simmons, R. F. (1962). Institute of Radio Engineers Transactions.

War Games. Ministry of Defence (Army).

War Games and Simulations. Ministry of Defence Library.
War Games with Models. Ministry of Defence (Army).

2.2.2 Magazines and Periodicals

There are an increasing number of magazines and periodicals devoted to gaming and simulation. Details of these publications, and others which have fairly regular coverage of the field, are given below.

Avalon Hill Intercontinental Kriegspiel Society European Region Newsletter
    22 Salisbury Road, Seaford, Sussex (Tucker, H.).
Airfix Magazine
    Surridge, Dawson & Co. (Publications), Publishing Dept, 136-142 New Kent Road, London S.E.1. Monthly. Annual subscription: £2.20.
Albion
    13 Gilmerston Court, Trumpington Road, Cambridge CB2 2HQ (Turnbull, D. J.). Bi-monthly. Annual subscription: 60p.
American Behavioural Scientist
    Sage Publications Inc., 275 South Beverly Drive, Beverly Hills, California 90212, U.S.A. (Greaser, C.) Bi-monthly. Annual subscription: $12.00.
Battleflag
    The Third Millenia Inc, 465 Woodland Hills, Philadelphia, Mississippi 39350. Monthly.
Behavioural Simulation Newsletter
    Political Science Dept, Peoples Avenue Complex, Building D, Rennselaer Polytechnic Institute, Troy, New York 12181, U.S.A. (Withed, M. H.).
Bellicus
    30 Plungington Road, Preston, Lancashire (Haven, W.).
Black Spot
    345 Livesey Branch Road, Blackburn BB2 4QJ (Pimley, L.).
Bristol War Games Society Magazine
    Bristol War Games Society
British Journal of Educational Studies
    Faber & Faber Ltd, 24 Russell Square, London W.C.1. Three per year. Annual subscription: £2.00.
Bulletin of Environmental Education
    Education Unit, Town and Country Planning Association, 17 Carlton House Terrace, London SW1Y 5AS (Fyson, A.).
Bushwacker
    5307 Carriage Court, Baltimore, Maryland 21229, U.S.A. (Davis, F.).
Casualty Simulation
    Casualties Union, 20 Grange Road, Newcastle-upon-Tyne NE4 9LD (Nicholson, M.). Quarterly. Annual subscription: 25p.

Chess
> Chess (Sutton Coldfield) Ltd, Sutton Coldfield, Birmingham. Bi-weekly.
> Annual subscription: £1.75.

Coffee Talk
> 657 Canterbury Road, Vermont, Victoria 3133, Australia. Monthly. Annual
> subscription: A$2.40.

Computer Bulletin
> British Computer Society, 23 Dorset Square, London N.W.1. Quarterly.
> Annual subscription: £2.62½.

Computer Journal
> British Computer Society, 23 Dorset Square, London N.W.1. Quarterly.
> Annual subscription: £4.50.

Computer Weekly
> Iliffe Technical Publications Ltd, Dorset House, Stamford Street, London
> S.E.1. Weekly.

Conflict
> Simulations Design Corporation, PO Box 19096, San Diego, California 92119.
> U.S.A. (Lombardy, D. and Zwicker, K.). Bi-monthly.

Il Corriere Diplomatico
> Via Vecchia di Barbaricina 20, 56100 Pisa, Italy. (Manfredi, E.).

Courier
> 13 Gilmerton Court, Trumpington Road, Cambridge CB2 2HQ (Turnbull,
> D. J.).

Der Krieg
> 8 Rusholme Road, London SW15 3JZ (Jeffrey, G.).

Diplophobia
> 12315 Judson Road, Wheaton, Maryland 20906, U.S.A. (Miller, D.).

Educational Development
> 6 Stewarts Way, Maunden, Bishops Stortford, Essex (Riley, A. J.). Twice per
> year. Annual subscription: 60p.

Ethil The Frog
> 17 Monmouth Road, Oxford OX1 4TD (Piggott, J.).

Gamers' Guide
> PO Box 255, Rockville Centre, New York 11571, U.S.A.

Games and Puzzles
> 19 Broadlands Road, PO Box 4, London N6 4DF (Levin, G. M.). Monthly.
> Annual subscription: £2.40.

Games and Toys
> 30-31 Knightrider Street, London E.C.4. Monthly. Annual subscription: £1.50.

Gamesletter
> 12315 Judson Road, Wheaton, Maryland 20906, U.S.A. (Miller, D.).

Gamesman
> 12315 Judson Road, Wheaton, Maryland 20906, U.S.A. (Miller, D.).

Grafeti
19 Doocot Road, St Andrews, Fife, Scotland (Yare, B.). Bi-weekly.
Annual subscription: 52p.

Grenadier
Cheltenham Wargamers Club, Cheltenham, Gloucestershire.

Harpers Sports and Games
Harper Trade Journals Ltd, Cousin House, 22 Cousin Lane, London E.C.4.
Weekly. Annual subscription: £1.25.

Harvard Business Review
Soldiers Field, Boston, Massachusetts 02163, U.S.A. Bi-monthly. Annual
subscription: $15.00

Hoosier Archives
Grand-Place 7, B-4280 Hannut, Belgium (Feron, M.).

Instructional Simulation Newsletter
Teaching Research Division, Oregon State System of Higher Education,
Monmouth, Oregon 97361, U.S.A. (Twelker, P.). Three per year. Free.

International Journal of Control
Taylor & Francis Ltd, Red Lion Court, Fleet Street, London E.C.4 Monthly.
Annual subscription: £16.25.

I.S.I. Learning Letter
Instructional Simulations Inc., 2147 University Avenue, St Paul, Minnesota
55104, U.S.A. (Klietsch, R. G.). Five per year. Annual subscription: $1.00.

Mad Policy
15 Crouch Oak Lane, Addlestone, Surrey (Walkerdine, R.).

Mathematical Pie
100 Burman Road, Shirley, Solihull, Warwickshire. Three per year. Annual
subscription: 10p.

Measurement and Control
20 Peel Street, London W.8. Monthly. Annual subscription: £12.

Midgard
Finches, 7 Cambridge Road, Beaconsfield, Buckinghamshire (Haven, W.).

Military Modelling
13-35 Bridge Street, Hemel Hempstead, Hertfordshire. Monthly. Annual
subscription: £2.25.

Miniature Warfare
1 Burnley Road, Stockwell, London S.W.9. Monthly. Annual subscription:
£3.15.

Model Engineer
13-35 Bridge Street, Hemel Hempstead, Hertfordshire. Fortnightly. Annual
subscription: £3.50.

Modern Teaching
657 Canterbury Road, Vermont, Victoria 3133, Australia (Williams, R.).
Three times per year. Annual subscription: A$2.00.

Moeshoeshoe
> Grand-Place 7, B-4280 Hannut, Belgium (Feron, M.).

Moves
> Duntay, 7 Alexander Drive, Timperley, Altrincham, Cheshire, WA15 6 NF (Watson, M.).

News
> Management Games Ltd, 70 Baker Street, Weybridge, Surrey. Occasional. Free.

Newsletter
> University of Utrecht, Dept of Sociology, PO Box 13015, Utrecht, Holland. Occasional. Free.

New Scientist
> New Science Publications Ltd, Cromwell House, Fulwood Place, High Holborn, London W.C.1. (Dixon, B.). Weekly. Annual subscription: £8.50.

Occasional Newsletter About Uses of Simulations and Games for Education and Training
> Project Simile Western Behavioural Sciences Institute, 1150 Silverado Road, La Jolla, California 92037, U.S.A. Three per year. Annual subscription: $5.00.

Operational Research Quarterly
> Pergamon Press Ltd, Headington Hill Hall, Oxford. Quarterly. Annual subscription: £5.00.

Panzerfaust
> PO Box 1123, Evansville, Indiana 47713, U.S.A. (Lowry, D.). Annual subscription: $5.00.

Political and Social Simulation Newsletter
> Haverford College, Haverford, Pennsylvania 19041, U.S.A. (Hsia, D.). Monthly. Annual subscription: $6.00.

Programmed Learning and Educational Technology
> Sweet & Maxwell Ltd, 11 New Fetter Lane, London EC4P 4EE (Unwin, D.). Bi-monthly. Annual subscription: £5.00.

SAGSET News
> 5 Errington, Moreton-in-Marsh, Gloucestershire (Teather, D.). Quarterly. Annual subscription: free to members.

School Services Newsletter
> Foreign Policy Association, 345 East 46th Street, New York, N.Y. 10017, U.S.A. Three per year.

Simulation
> Simulation Councils Inc., PO Box 8248, San Diego, California (McLeod, J.). Monthly. Annual subscription: $28.00.

Simulation and Games: An International Journal of Theory, Design and Research
> Sage Publications Inc., 275 South Beverly Drive, Beverly Hills, California 90212, U.S.A. (Inbar, M.). Quarterly. Annual subscription: $15.00.

Slingshot
757 Pershore Road, Birmingham B29 7NY (Barker, P.). Bi-monthly. Annual subscription: £1.50.

Strategy and Tactics
Duntay, 7 Alexander Drive, Timperley, Altrincham, Cheshire, WA15 6NF (Watson, M.).

Strategy and Tactics Guide
Simulations Publications (Campion, M. and Phillies, G.). Quarterly. Annual subscription: $4.00.

Subbuteo Soccer News
Subbuteo Sports Games Ltd, Langton Green, Tunbridge Wells, Kent (Erik, G.).

Times Educational Supplement
Times Newspapers Co. Ltd, Printing House Square, London EC4. Weekly. Annual subscription: £3.12.

The Diplomacy Backstabber
(Levin, G.).

The Official Gazette of the Grand Duchy of Pfennig-Halb-Pfennig
PO Box 52, Ralston, Alberta TOJ 2NO, Canada (McCallum, J.).

The Setwork Network
Hog End, Bloxham, Banbury, Oxfordshire (Aylmer, R. G.). Three per year.

Visual Education
National Committee for Audio-Visual Aids in Education, 33 Queen Anne Street, London W.1.

VOSA News
Voluntary Overseas Service Association, 26 Museum Chambers, Little Russell Street, London WC1A 2JA (Parry, A.).

War Bulletin
Finches, 7 Cambridge Road, Beaconsfield, Buckinghamshire (Patterson, H.). Monthly. Annual subscription: 48p.

Wargamers Newsletter
69 Hill Lane, Southampton, Hampshire, SO1 5AD (Featherston, D.). Monthly. Annual subscription: £1.80.

Wff'n Proof Newsletter
Wff'n Proof Inc., PO Box 71, New Haven, Connecticut 06501, U.S.A. (Patmore, R. C.). Quarterly. Annual Subscription: $1.00.

2.2.3 Books

ABT, C. C. (1966) *Games for Learning.* Stevensville, Educational Services Inc.

ABT, C. C. (1965) *Heuristic Games for Secondary Schools.* Cambridge, Mass., Abt Associates Inc.

ABT, C. C. (1970) *Serious Games.* New York, Viking Press.

ABT, C. C. (1966) *The URB-COIN Urban Insurgency Game.* Washington, D.C., U.S. Dept of Defence.

ABT, C. C. (1967) *Ten Steps in Game Design.* Cambridge, Mass., Abt Associates Inc.

ABT, C. C. *Training Games.* Cambridge, Mass., Abt Associates Inc.

ABT, C. C. (1965) *Game Learning and Disadvantaged Groups.* Cambridge, Mass., Abt Associates Inc.

ABT, C. C. (1966) *COCON—Counterconspiracy (Politica): The Development of a Simulation of Internal National Conflict under Revolutionary Combat Conditions.* Cambridge, Mass., Abt Associates Inc.

ABT, C. C. (1966) *Six Demonstrations of the AGILE/COIN Game.* Cambridge, Mass., Abt Associates Inc.

ALLEN, L. E. and CALDWELL, M. E. (Ed.) (1965) *Communication Sciences and Law: Reflections from the Jurimetric Conference.* Indianapolis, Bobbs Merrill Co.

ALLEN, L. E. and CALDWELL, M. E. (1957) *Top Management Decision Simulation.* New York, American Management Association.

ALLEN, L. E. and CALDWELL, M. E. (1961) *Simulation and Gaming: A Symposium.* New York, American Management Association.

ANDERSON, L. F. (1964) *Combining Simulation and Case Studies in the Teaching of American Foreign Policy.* Evanston, Northwestern University.

ARGYRIS, C. *Role-Playing in Action.* Ithaca, Cornell University.

ARMSTRONG, R. H. R. and TAYLOR, J. L. (Ed.) (1971) *Feedback on Instructional Simulation Systems.* Cambridge, Cambridge Institute of Education.

ARMSTRONG, R. H. R; and TAYLOR, J. L. (Ed.) (1970) *Instructional Simulation Systems in Higher Education.* Cambridge, Cambridge Institute of Education.

ARROW, K. (1951) *Social Choice and Individual Values.* New York, Wiley.

AVEDON, E. M. and SUTTON-SMITH, B. (1971) *The Study of Games.* New York, Wiley.

BABB, E. M. and EISGRUBBER, L. M. (1966) *Management Games for Teaching and Research.* Chicago, Educational Methods Inc.

BALDET, M. (1961) *Lead Soldiers and Figurines.*

BALLANTYNE, F. P. (1954) *A Two Machine Gun Duel with the Bomber Turret Vulnerable.* Santa Monica, Rand Corporation.

BAMFORD, H. E. (1956) *The Use of Training Aids in Conceptual Training.* Pittsburgh, American Institute for Research.

BARD, B. (1957) *Making and Collecting Military Miniatures.*

BASIL, D. C. *et al.* (1965) *Executive Decision Making Through Simulation.* Englewood Cliffs, Prentice-Hall.

BELLMAN, R. E. *et al.* (1949) *Application of the Theory of Games to Identification of Friend and Foe.* Santa Monica, Rand Corporation.

BELLMAN, R. E. and BLACKWELL, D. (1949) *An Example of Bluffing with Pure Strategies.* Santa Monica, Rand Corporation.

BELZER, R. L. (1950) *Some Remarks on Best Strategies.* Santa Monica, Rand Corporation.

BEMERS-LEE, C. M. (Ed.) (1965) Models for Decision. London. E.U.P.

BENGOUGH, H. M. (1891) *Illustrations of Field Exercises of the Three Arms, of Exercises in Minor Tactics, and War Games.*

BERGE, C. and GHOUILLA-HOURI, A. (1965) *Programming, Games and Transportation Networks.* London, Methuen.

BERNE, E. (1964) *Games People Play.* Harmondsworth, Penguin.

BERKOVITZ, L. D. (1960) *The Implications of Some Game Theoretic Analyses for War Gaming.* Santa Monica, Rand Corporation.

BLACKWELL, D. and GIRSHICK, M. A. (1954) *Theory of Games and Statistical Decisions.* New York, Wiley.

BLAQUIERE, A. *et al.* (1969) *Quantitative and Qualitative Games.* New York, Academic Press.

BLOOMFIELD, L. P. (1961) *The Political Exercise: A Progress Report.*

BLUM, P. (1964) *Military Miniatures.* London, Hamlyn.

BLUM, P. (1971) *Model Soldiers: A Basic Guide to Painting, Animating and Converting.* London, Arms and Armour Press.

BONINI, C. (1962) *Simulation of Information and Decision Systems in the Firm.* Ithaca, N.Y. Carnegie Institute of Technology.

BOOCOCK, S. and SCHILD, E. O. (Ed.) (1968) *Simulation Games in Learning.* Beverly Hills, Sage Publications Inc.

BORKO, H. (Ed.) (1962) *Computer Applications in the Behavioural Sciences.* Englewood Cliffs, Prentice-Hall.

BROOM, H. N. (1969) *Business Policy and Strategic Action: Text, Cases and Management Game.* Englewood Cliffs, Prentice-Hall.

BROM, J. R. (1955) *Narrative Description of an Analytic Theater Air Ground System.* Santa Monica, Rand Corporation.

BROSSMAN, M. W. *War Gaming.* McLean, Va., Research Analysis Corporation.

BUCHLER, I. R. (Ed.) (1966) *Applications of the Theory of Games.* Austin, University of Texas.

BUCHLER, I. R. and NUTUNI, H. G. (Ed.) (1969) *Game Theory in the Behavioural Sciences.*

BURTON, J. W. (1965) *Simulation in the Teaching of International Relations.* London, University College.

BURTON, J. W. (1968) *Systems, States, Diplomacy and Rules.* London, University College.

BUSHNELL, D. D. (1965) *The Automation of School Information Systems.* Washington, D.C., Dept of Audio-Visual Instruction.

CALDER, N. (1969) *The Environment Game.* London, Panther.

CALLOIS, R. (1963) *Man, Play and Games.* London, Thames & Hudson.

CAMPBELL, A. *et al.* (1954) *The Voter Decides.* Evanston, Ron Peterson.

CAMPBELL, E. *et al.* (1966) *Equality of Educational Opportunity.* Washington, D.C., U.S. Office of Education.

CAMPBELL, E. *et al.* (1965) *A Digital Computer Simulation of an Artillery Barrage for the Evaluation of Area Kill Weapons.* Montreal, Dept of National Defence.

CANDLER, J. C. (1964) *Miniature War Games 'du temps du Napolëon'.*

CAREY, E. G. and GILBRETH, F. B. (1948) *Cheaper by the Dozen.* New York, Gross & Dunlop Inc.

CARLSON, E. (1968) *Learning Through Games.* Washington, D.C., Public Affairs Press.

CARLSON, E. (1968) *Further Relations Between Game Theory and Eigen Systems.* Carnegie-Mellon University.

CHESLER, M. and FOX, R. (1967) *Role-Playing in the Classroom.* Henley, Science Research Associates.

CHESLER, M. *et al.* (1967) *Problem-Solving to Improve Classroom Learning.* Henley, Science Research Associates.

CHORAFAS, D. N. (1965) *Systems and Simulation.* New York, Academic Press.

CHU, Y. (1967) *Digital Analog Simulation Techniques.* Baltimore, Maryland University.

CHURCHILL, N. C. *et al.* (1964) *Auditing, Management Games and Accounting Education.* Homewood, R. D. Irwin Inc.

CLARKE, J. C. (1872) *Kriegspiel or Game of War.*

COGHILL, M. A. (1971) *Games and Simulations in Industrial and Labour Relations Training.* Ithaca, Cornwell University.

COGSWELL, J. F. and ESTAVAN, D. P. (1965) *Computer Simulation of a Counselor in Student Appraisal and in the Educational Planning Interview.* Santa Monica, System Development Corporation.

COGSWELL, J. F. *et al.* (1964) *New Solutions to Implementing Instructional Media Through Analysis and Simulation of School Organization.* Santa Monica, System Development Corporation.

COHEN, K. J. *et al.* (1964) *The Carnegie Management Game: An Experiment in Business Education.* Homewood, Richard D. Irwin Inc.

COHEN, N. D. (1966) *An Attack Defence Game with Matrix Strategies.* Santa Monica, Rand Corporation.

COLEMAN, J. S. (1965) *Adolescents and the Schools.* New York, Basic Books.

COLEMAN, J. S. (1964) *Introduction to Mathematical Sociology.* New York, Free Press.

CRAWFORD, M. P. (1967) *Simulation in Education and Training.* Alexandria, Human Resources Research Office.

CROOK, H. T. (1888) *War Game Maps.*

CRUICKSHANK, D. R. and BROADBENT, F. W. (1969) *Simulation in Preparing Educational Personnel.* Washington, ERIC Centre for Teacher Education.

CRUICKSHANK, D. R. and BROADBENT, F. W. (1968) *The Simulation and Analysis of Problems of Beginning Teachers.* Washington, D.C., U.S. Dept of Health, Education & Welfare.

CULBERTSON, J. A. and HENCLEY, S. P. (1963) *Educational Research, New Perspectives.* Danville, Interstate Printers & Publishers.

CULBERTSON, J. A. and COFFIELD, W. H. (1960) *Simulation in Administrative Training.* Columbus, University Council for Educational Administration.

DALE, A. G. (1963) *Simulation Training for Small Business Executive Development.* Austin, University of Texas.

DALE, A. G. and KLASSON, C. B. (1964) *Business Gaming: A Survey of American Collegiate Schools of Business.* Austin, University of Texas.

DALKEY, N. C. *Simulation of Military Conflict.* Santa Monica, Rand Corporation.

DALKEY, N. C. (1966) *STROP: A Strategic Planning Model.* Santa Monica, Rand Corporation.

DALKEY, N. C. (1964) *Solvable Nuclear War Models.* Santa Monica, Rand Corporation.

DALTON, R. *et al.* (1972) *Simulation Games in Geography.* New York, Macmillan.

DARDEN, W. R. and LUCAS, W. M. (1969) *The Decision Making Game: An Integrated Operations Management Simulation.* New York, Appleton-Century-Crofts.

DAWSON, G. G. (Ed.) (1967) *Economic Education Experiences of Enterprising Teachers.* New York, Joint Council on Economic Education.

DE MILLE, R. (1967) *Put Your Mother on the Ceiling.* New York, Walker.

DE WEERD, H. A. (1967) *Political-Military Scenarios.* Santa Monica, Rand Corporation.

DILL, W. R. *et al.* (Ed.) (1961) *Proceedings of the Conference on Business Games as Teaching Devices.* New Orleans, Tulane University.

DRESHER, M. *et al.* (Ed.) (1964) *Advances in Game Theory.*

DRESHER, M. (1956) *The Theory of Games of Strategy.* Santa Monica, Rand Corporation.

DRESHER, M. (1959) *Some Military Applications of the Theory of Games.* Santa Monica, Rand Corporation.

DRESHER, M. (1961) *Games of Strategy.* Englewood Cliffs, Prentice-Hall.

DUKE, R. D. (1965) *Gaming Urban Systems Planning.* Washington, D.C., American Society of Planning Officials.

DUKE, R. D. (1964) *Gaming Simulation in Urban Research.* Ann Arbor, Michigan State University.

DUNN, P. (1970) *Sea Battle Games.* Aylesbury, Model and Allied Publications.

DUNN, W. R. and HOLROYD, C. (Ed.) (1969) *Aspects of Educational Technology II.* London, Methuen.

DURBIN, E. P. (1966) *TARLOG: A Differential Ground Combat Model.* Santa Monica, Rand Corporation.

EDDY, A. G. and HEWETT, P. C. (1961) *Player Participation Gaming in Limited War Applications.* Omega Technical Operations.

EDMUNDSON, H. P. (1958) *The Distribution of Radial Error and its Statistical Application in War Gaming.* Santa Monica, Rand Corporation.

ELLIOTT BROTHERS LTD. (1966) *A Digital Simulation Language for the Elliot 503 Digital Computers (SLANG),* London, Elliott Brothers (London) Ltd.

EMSHOFF, J. R. and SISSON, R. L. (1971) *Design and Use of Computer Simulation Models.* London, Collier-Macmillan.

EMSHOFF, J. R. and SISSON, R. L. (1968) *The Role of Computer Simulation in Education and Training.* Austin, Entelek Inc.

EMSHOFF, J. R. and SISSON, R. L. (1963) *Control and Simulation Language.* London, Esso Petroleum Co.

EVANS, G. W. *et al.* (1967) *Simulation Using Digital Computers.* Englewood Cliffs, Prentice-Hall.

FADEN, B. R. (Ed.) (1959) *Proceedings of the National Symposium on Management Games.* Wichita, University of Kansas.

FAIRHEAD, J. N., PUGH, D. S. and WILLIAMS, W. J. (1965) *Exercises in Business Decisions.* London, E.U.P.

FARMER, J. A. and COLLCUT, R. H. (1965) *Experience of Digital Simulations in a Large O.R. Group.* London, British Iron and Steel Research Association.

FATTU, N. and ELAM, S. (Ed.) (1965) *Simulation Models for Education.* Bloomington, Phi Delta Kappa.

FEATHERSTONE, D. F. (1965) *Naval War Games.* London, Stanley Paul Co.

FEATHERSTONE, D. F. (1970) *War Game Campaigns.* London, Stanley Paul Co.

FEATHERSTONE, D. F. (1962) *War Games.* London, Stanley Paul Co.

FEATHERSTONE, D. F. (1969) *Handbook for Model Soldier Collectors.* London, Kaye and Ward.

FEATHERSTONE, D. F. (1963) *Tackle Model Soldiers This Way.*

FEATHERSTONE, D. F. (1969) *Advanced War Games.* London, Stanley Paul Co.

FEATHERSTONE, D. F. (1970) *Battles with Model Soldiers.* Newton Abbott, David and Charles.

FEATHERSTONE, D. F. (1966) *Air War Games.* London, Stanley Paul Co.

FEIGENBAUM, E. A. and FELDMAN, J. (Ed.) (1963) *Computers and Thought.* New York, McGraw-Hill.

FELDT, A. G. (1965) *The Cornell Land Use Game.* Ann Arbor, Michigan State University.

FULKERSON, D. R. and JOHNSON, S. M. (1957) *A Tactical Air Game.* Santa Monica, Rand Corporation.

FULMER, J. L. Business Simulation Games. Cincinnati, Southwestern Publishing Co.

GAGNÉ, R. M. (1962) *The Conditions of Learning.* New York, Holt, Rinehart & Winston.

GAGNÉ, R. M. *et al.* (1962) *Psychological Principles in System Development.* New York, Holt, Rinehart & Winston.

GALLER, B. A. (1962) *The Language of Computers.* New York, McGraw-Hill.

GAMSON, W. A. (1966) *Simsoc: A Manual for Participants.* Ann Arbor, Campus Publishers.

GARRETT, J. G. (1959) *Model Soldiers: A Collector's Guide.* London, Seeley.

GEISLER, M. A. (1959) *The Simulation of a Large-Scale Military Activity.* Santa Monica, Rand Corporation.

GIBSON, J. J. (Ed.) (1947) *Motion Picture Testing and Research.* Washington, D.C., Army Air Forces.

GIFFIN, S. F. (1965) *The Crisis Game: Simulating International Conflict.* New York, Doubleday.

GILLET, W. (1872) *The Game of War.*

GLASER, R. (Ed.) (1965) *Training Research and Education.* New York, Wiley.

GLAZIER, R. (1967) *Development of Foreign Policy Simulation Exercises: A Manual.* New York, Foreign Policy Association.

GLAZIER, R. (1969) *How to Design Educational Games.* Cambridge, Mass., Abt Associates Inc.

GLICKSBERG, I. and GROSS, O. (1951) *Continuous Games with Given Strategies.* Santa Monica, Rand Corporation.

GORDON, A. K. (1970) *Games for Growth.* Chicago, Science Research Associates.

GORDON, G. (1969) *System Simulation.* Englewood Cliffs, Prentice-Hall.

GRAHAM, R. G. and GRAY, C. F. (1969) *Business Games Handbook.* New York, American Management Association.

GRANT, C. (1971) *The War Game.* London, A. & C. Black.

GRANT, C. (1970) *Battle! Practical Wargaming.* Aylesbury, Model and Allied Publications.

GREEN, B. F. (1963) *Digital Computers in Research: An Introduction for Behavioural and Social Scientists.* New York, McGraw-Hill.

GREEN, J. R. and SISSON, R. L. (1959) *Dynamic Management Decision Games.* New York, Wiley.

GREENBERGER, M. *et al.* (1965) *On Line Computation and Simulation: The OPS-3 System.* Cambridge, Mass., MIT Press.

GREENE, T. E. (1960) *The 'Contextual Study' Method as a Device for Studying Limited War Strategies.* Santa Monica, Rand Corporation.

GREENLAW, P. S. *et al.* (1962) *Business Simulation in Industrial and University Education.* Englewood Cliffs, Prentice-Hall.

GUETZKOW, H. *et al.* (1963) *Simulation in International Relations: Developments for Research and Teaching.* Englewood Cliffs, Prentice-Hall.

GUETZKOW, H. *et al.* (1962) *Simulation in Social Science: Readings.* Englewood Cliffs, Prentice-Hall.

GUETZKOW, H. and KOTLER, P. (Ed.) (1970) *Simulation in Social and Administrative Science.* Englewood Cliffs, Prentice-Hall.

HANTZES, H. N. *et al.* (1959) *Development of Intelligence Requirements Through the Interrogation of War Game Players.* Baltimore, John Hopkins University.
HARE, P. A. (Ed.) (1955) *Small Groups.* New York, Knopf.
HARMAN, H. H. (Ed.) (1961) *Simulation: A Survey.* Los Angeles, Western Joint Computer Conference.
HARRIS, H. E. D. (1969) *How to Go Collecting Model Soldiers.* London, Patrick Stephens.
HARRIS, H. E. D. (1962) *Model Soldiers.* London, Wiedenfeldt and Nicholson.
HARVER, O. J. (Ed.) (1966) *Flexibility, Adaptability and Creativity.* New York, Springer Inc.
HAUSNER, M. (1952) *Optimal Strategies in Games of Survival.* Santa Monica, Rand Corporation.
HAYWOOD, A. G. (1889) *Map Manoeuvres: A Simple Account of the War Game.*
HAYWOOD, A. G. (1896) *Kriegspiel as Applied to Field Manoeuvres.*
HAYWOOD, O. G. (1951) *Military Doctrine of Decision and the Von Neumann Theory of Games.* Santa Monica, Rand Corporation.
HELMER, O. (1960) *Strategic Gaming.* Santa Monica, Rand Corporation.
HEMPHILL, J. K. *et al.* (1962) *Administrative Performance and Personality: A Study of the Principal in a Simulated Elementary School.* New York, Columbia University.
HENSHAW, R. C. and JACKSON, J. R. (1966) *The Executive Game.* Homewood, R. D. Irwin.
HERMAN, C. F. and HERMAN, M. (1962) *On the Possible Use of Historical Data for Validation Study of Inter-Nation Simulation.* Evanston, Northwestern University.
HERRON, L. W. (1960) *Executive Action Simulation.* Englewood Cliffs, Prentice-Hall.
HIRSCH, W. Z. (Ed.) (1967) *Inventing Education of the Future.* San Francisco, Chandler Publishing Co.
HOGGAT, A. C. and BALDERSTON, F. E. (Ed.) *Symposium on Simulation Models.* Cincinnati, Southwestern Publishing Co.
HOLLINGDALE, S. H. (Ed.) (1967) *Digital Simulation in Operational Research.* Paris, North Atlantic Treaty Organisation.
HUIZINGA, J. (1970) *Homo Ludens.* London, Paladin.
HUNT, E. B. (1962) *Concept Learning: An Information Processing Problem.* New York, Wiley.
HUNT, E. B. and HOOLAND, C. I. (Ed.) (1960) *Programming a Model of Human Concept Formation.* Western Joint Computer Conference.
IMMANUEL, (1967) *The Regimental War Game.*
IMMEGART, G. C. *Guides for the Preparation of Instructional Case Materials in Educational Administration.* Columbia, Universities Council for Educational Administration.
INBAR, M. *et al.* (1970) *Developing Social Simulations.* New York, Free Press.

ISAACS, R. P. (1951) *Games of Pursuit.* Santa Monica, Rand Corporation.

ISAACS, R. P. (1965) *Differential Games: A Mathematical Theory with Applications to Warfare and Pursuit, Control and Optimization.* Santa Monica, Rand Corporation.

JAFFEE, C. L. (1968) *Problems in Supervision.* Reading, Mass., Addison-Wesley.

JAMES, M. L. *et al.* (1966) *Analog Computer Simulation of Engineering Systems.*

JENNESS, R. R. (1965) *Analog Computation and Simulation: Laboratory Approach.*

KAHN, H. and MANN, I. (1957) *War Gaming.* Santa Monica, Rand Corporation.

KAPLAN, M. A. (Ed.) (1968) *New Approaches to International Relations.* St Martins Press.

KARPLUS, W. J. (1958) *Analog Simulation: Solution of Field Problems.* New York, McGraw-Hill.

KAUFMANN, A. (1967) *Graphs, Dynamic Programming and Finite Games.* New York, Academic Press.

KAZEMIER, B. H. and VUYSJE, D. (Ed.) (1963) *The Concept and the Role of the Model in Mathematics and Natural and Social Sciences.* New York, Gordon & Breach.

KEIGHER, R. M. (1960) *War Game Evaluation Methods.* Omega Technical Operations.

KENNEDY, J. L. (1963) *Simulation Studies of Organization.* Princeton University Press.

KENT, G. (1968) *The Effects of Threats.*

KERSH, B. Y. (1963) *Classroom Simulation: A New Dimension in Teacher Education.* Monmouth, Oregon System of Higher Education.

KERSH, B. Y. (1965) *Classroom Simulation: Further Studies on Dimensions of Realism.* Monmouth, Oregon System of Higher Education.

KIBBEE, J. M. *et al.* (1961) *Management Games: A New Technique for Executive Development.* New York, Reinhold.

KITCHENER, F. W. (1895) *Rules for War Games on Maps and Tactical Models.* Simla, Government Central Printing Office.

KLEIN, A. F. (1966) *Role-Playing in Leadership Training and Group Problem-Solving.* New York, Association Press.

KLIETSCH, R. G. (1968) *An Introduction to Learning Games and Instructional Simulations.* St Paul, Instructional Simulations Co.

KNUDSON, H. R. (1963) *Human Elements in Administration.* New York, Holt, Rinehart & Winston.

KOHN, C. F. (1964) *Selected Classroom Experiences: High School Geography Project.* Normal, National Council for Geographic Education.

KORN, G. A. (1966) *Random-Process Simulation and Measurements.* New York, McGraw-Hill.

KORNS, M. F. (1966) *The Modern War in Miniature: A Statistical Analysis of the Period 1939 to 1945.*

KUGEL, P. (1961) *Display Systems for Digital Simulations.* Technical Operations Inc.

KUHN, H. W. and TUCKER, A. W. (Ed.) (1953) *Contributions to the Theory of Games.* Princeton, Princeton University Press.

KURFMAN, D. G. and PHILLIPS, I. M. *Teaching Procedures for the New Social Studies Using Simulation to Involve Students.*

KURUE, M. I. (1962) *Human Factors in War Gaming.* Technical Operations Inc.

LEGGETT, T. (1966) *Shogi: Japan's Game of Strategy.* Englewood Cliffs, Prentice-Hall.

LETH-ESPENSEN, J. and CLAUSSEN, L. K. (1963) *Computer Technique in Simulation of Air Battles.* Versailles, Supreme Headquarters Allied Powers Europe.

LEVIN, R. I. and DES JARDINS, R. B. (1970) *Theory of Games and Strategies.*

LIFE, A. and PUGH, D. (Ed.) (1965) *Business Exercises: Some Developments.* Oxford, Blackwell.

LIVERMORE, W. R. (1882) *The American Kriegspiel.*

LOPEZ, F. M. (1966) *Evaluating Executive Decision Making: The In-Basket Technique.* New York, American Management Association.

LOVELUCK, C. *How to Conduct a Business Game.* London, B. L. Industrial Training Aids Ltd.

LOVELUCK, C. (1969) *Notes on the Construction, Operation and Evaluation of Management Games.* Weybridge, Management Games Ltd.

LUCE, R. D. and RAIFFA, H. (1957) *Games and Decisions.* New York, Wiley.

LUND, V. E. (1966) *Evaluation of Simulation Techniques to Teach Dental Office Emergencies.* Monmouth, Oregon State System of Higher Education.

MCDONALD, J. (1950) *Strategy in Poker, Business and War.*

MCKENNEY, J. L. (1967) *Simulation Gaming for Management Development.* Cambridge, Mass., Harvard University.

MCLEOD, J. (Ed.) (1968) *Simulation: The Dynamic Modeling of Ideas and Systems with Computers.*

MACPHERSON, W. G. (1911) *The Instruction of Medical Officers in the Austro-Hungarian Army by Means of War Games.*

MACUNOVICH, D. (1967) *Planning Games.* London, Transport Research and Computation Co. Ltd.

MANN, A. P. and BRUNSTRÖM, K. (Ed.) (1969) *Aspects of Educational Technology III.* London, Pitman.

MARTIN, F. F. (1968) *Computer Modeling and Simulation.* New York, Wiley.

MEIER, R. C. et al. (1969) *Simulation in Business and Economics.* Englewood Cliffs, Prentice-Hall.

MENSCH, A. (Ed.) (1966) *Theory of Games: Techniques and Applications.* Toulon, North Atlantic Treaty Organisation.

MENSCH, A. (1960) *Computer Simulation of Vehicle Motion in Three Dimensions.* Lansing, Michigan University.

MIDDLETON, (1873) *Explanation and Application of the English Rules for the War Game.*

MILNOR, J. and SHAPLEY, L. S. (1956) *Games of Survival.* Santa Monica, Rand Corporation.

MIZE, J. H. and COX, J. G. (1968) *Essentials of Simulation.* Englewood Cliffs, Prentice-Hall.

MONROE, M. W. (1961) *Games as Teaching Tools: An Examination of the Community Land Use Game.* Ithaca, Cornell University.

MORSCHAUER, J. (1962) *How to Play War Games in Miniature.* New York, Walker & Co.

NAYLOR, R. H. *et al.* (1966) *Computer Simulation Techniques.* New York, Wiley.

NESBITT, W. A. (1966) *New Dimensions: Simulation Games for the Social Studies Classroom.* New York, Foreign Policy Association.

NEUMANN, J. VON and MORGENSTERN, O. (1964) *Theory of Games and Economic Behaviour.* New York, Wiley.

NEUMANN, J. VON and MORGENSTERN, O. (1968) *Negotiations and Decisions In a Politics Game.* New York State University.

NEWELL, A. (Ed.) (1961) *Information Processing Language V Manual.* Englewood Cliffs, Prentice-Hall.

NEWELL, A. and SIMON, H. A. (1959) *The Simulation of Human Thought.* Santa Monica, Rand Corporation.

NICHOLSON, J. B. R. (Ed.) (1967) *Model Soldiers.* Belmont-Maitland.

NICOLLIER, J. (1967) *Collecting Toy Soldiers.* Englewood Cliffs, Prentice-Hall.

NOEL, R. C. (1960) *Evolution of the Inter-Nation Simulation.* Evanston, Northwestern University.

NOEL, R. C. (1966) *Chance-Constrained Games with Partially Controllable Strategies.* Evanson, Northwestern University.

ORCUTT, G. H. *et al.* (1961) *Microanalysis of Socio-Economic Systems: A Simulation Study.* New York, Harper & Row.

OWEN, G. (1968) *Game Theory.* London, W. B. Saunders.

PATTISON, W. D. (1963) *Means of Instruction: An Alphabetical Selection.* Boulder, High School Geography Project.

PAXSON, E. W. (1963) *War Gaming.* Santa Monica, Rand Corporation.

PEISAKOFF, M. P. (1952) *More on Games of Survival.* Santa Monica, Rand Corporation.

PESTON, M. and CODDINGTON, A. (1967) *The Elementary Ideas of Game Theory.* London, H.M.S.O.

PIGORS, P. and PIGORS, F. (1961) *Case Method in Human Relations: The Incident Process.* New York, McGraw-Hill.

PLATIS, M. E. (Ed.) *Plus: A Handbook for Teachers of Elementary Arithmetic.* Stevensville, Educational Service Inc.

POOL, I. S. *et al.* (1964) *Candidates, Issues and Strategies.* Cambridge, Mass. M.I.T. Press.

POOL, I. S. *et al.* (1963) *The Kernel of a Cooperative Game.* Princeton, Princeton University.

POOL, I. S. *et al.* (1967) *'Simple' Stability of General n-Person Games.* Princeton, Princeton University.

POST OFFICE RESEARCH DEPARTMENT (1965) *Computer Processing of Signals with Particular Reference to Simulation of Electronic Filter Networks.* London, Post Office Corporation.

PRITSKÉR, A. A. B. and KÍVIAT, P. J. (1969) *Simulation with GASP II: A Fortran Based Simulation Language.*

PRITSKÉR, A. A. B. and KÍVIAT, P. J. (1968) *The Tactical and Negotiations Game: A Simulation of Local Conflict, an Analysis of some Psychopolitical and Applied Implications of TNG Simulation Research.* Lajayette, Purdue University.

PRITSKÉR, A. A. B and KÍVIAT, P. J. (1952) *The Uses and Limitations of Mathematical Models, Game Theory and Systems Analysis in Planning and Problem Solution.* Santa Monica, Rand Corporation.

PRITSKÉR, A. A. B. and KÍVIAT, P. J. (1959) *Systems Analysis and Education.* Santa Monica, Rand Corporation.

PRITSKÉR, A. A. B. and KÍVIAT, P. J. (1952) *Introduction to the Theory of Games.* Santa Monica, Rand Corporation.

PRITSKÉR, A. A. B. and KÍVIAT, P. J. (1962) *Continuous Games with Perfect Information.* Santa Monica, Rand Corporation.

PRITSKÉR, A. A. B. and KÍVIAT, P. J. (1963) *An Approach to the Study of a Developing Economy by Operational Gaming.* Santa Monica, Rand Corporation.

PRITSKÉR, A. A. B. and KÍVIAT, P. J. (1964) *Games and Simulations.* Santa Monica, Rand Corporation.

PRITSKÉR, A. A. B. and KÍVIAT, P. J. (1967) *Utility Comparison and the Theory of Games.* Santa Monica, Rand Corporation.

PRITSKÉR, A. A. B. and KÍVIAT, P. J. (1967) *The 'Value of the Game' as a Tool in Theoretical Economics.* Santa Monica, Rand Corporation.

PRITSKÉR, A. A. B. and KÍVIAT, P. J. (1969) *Proper Raising Points in a Generalization of Backgammon.* Santa Monica, Rand Corporation.

PRITSKÉR, A. A. B. and KÍVIAT, P. J. (1951) *Continuous Games with Given Strategies.* Santa Monica, Rand Corporation.

PRITSKÉR, A. A. B. and KÍVIAT, P. J. (1951) *Military Doctrine of Decision and the Von Neuman Theory of Games.* Santa Monica, Rand Corporation.

PRITSKÉR, A. A. B. and KÍVIAT, P. J. (1952) *Optimal Strategies in Games of Survival.* Santa Monica, Rand Corporation.

PRITSKÉR, A. A. B. and KÍVIAT, P. J. (1963) *Some Thoughts on the Theory of Cooperative Games.* Santa Monica, Rand Corporation.

PRITSKÉR, A. A. B. and KÍVIAT, P. J. (1963) *Compound Simple Games II: Some General Composition Games.* Santa Monica, Rand Corporation.

PRITSKÉR, A. A. B. and KÍVIAT, P. J. (1965) *Notes on n-Person Games VII: Cores of Convex Games.* Santa Monica, Rand Corporation.

PRITSKÉR, A. A. B. and KÍVIAT, P. J. (1967) *Pure Competition, Coalition Power and Fair Division.* Santa Monica, Rand Corporation.

PRITSKÉR, A. A. B. and KÍVIAT, P. J. (1967) *Games with Unique Solutions which are Non-Convex.* Santa Monica, Rand Corporation.

PRITSKÉR, A. A. B. and KÍVIAT, P. J. (1967) *The Kernel and Bargaining Set for Convex Games.* Santa Monica, Rand Corporation.

PRITSKÉR, A. A. B. and KÍVIAT, P. J. (1967) *Compound Simple Games 3: On Committees.* Santa Monica, Rand Corporation.

PRITSKÉR, A. A. B. and KÍVIAT, P. J. (1968) *Values of Non-Atomic Games Part I: The Axiomatic Approach.* Santa Monica, Rand Corporation.

PRITSKÉR, A. A. B. and KÍVIAT, P. J. (1967) *The Game with No Solution.* Santa Monica, Rand Corporation.

PRITSKÉR, A. A. B. and KÍVIAT, P. J. (1968) *The Proof that a Game may not have a Solution.* Santa Monica, Rand Corporation.

PRITSKÉR, A. A. B. and KÍVIAT, P. J. (1968) *On Solutions for n-Person Games.* Santa Monica, Rand Corporation.

PRITSKÉR, A. A. B. and KÍVIAT, P. J. (1968) *On Market Games.* Santa Monica, Rand Corporation.

PRITSKÉR, A. A. B. and KÍVIAT, P. J. (1970) *Game Theory and Politics: Recent Soviet Views.* Santa Monica, Rand Corporation.

PRITSKÉR, A. A. B. and KÍVIAT, P. J. (1969) *Values of Non-Atomic Games Part 2: The Random Order Approach.* Santa Monica, Rand Corporation.

PRITSKÉR, A. A. B. and KÍVIAT, P. J. (1969) *Voting, or a Price System in a Competitive Market Structure.* Santa Monica, Rand Corporation.

PRITSKÉR, A. A. B. and KÍVIAT, P. J. (1969) *Games of Status.* Santa Monica, Rand Corporation.

PRITSKÉR, A. A. B. and KÍVIAT, P. J. (1969) *A Penetration Game Model with Homing but no Counting for the Defence.* Santa Monica, Rand Corporation.

PRITSKÉR, A. A. B. and KÍVIAT, P. J. (1963) *Simulating with SIMSCRIPT.* Santa Monica, Rand Corporation.

PRITSKÉR, A. A. B. and KÍVIAT, P. J. (1964) *GASP: A General Simulation Program.* Santa Monica, Rand Corporation.

PRITSKÉR, A. A. B. and KÍVIAT, P. J. (1964) *Computer Simulation of Human Behaviour.* Santa Monica, Rand Corporation.

PRITSKÉR, A. A. B. and KÍVIAT, P. J. (1965) *Man Machine Simulation Experience.* Santa Monica, Rand Corporation.

PRITSKÉR, A. A. B. and KÍVIAT, P. J. (1966) *Development of New Digital Simulation Languages.* Santa Monica, Rand Corporation.

PRITSKÉR, A. A. B. and KÍVIAT, P. J. (1966) *Simulation Language Report Generators.* Santa Monica, Rand Corporation.

PRITSKÉR, A. A. B. and KÍVIAT, P. J. (1966) *Development of Discrete Digital Simulation Languages.* Santa Monica, Rand Corporation.

PRITSKÉR, A. A. B. and KÍVIAT, P. J. (1969) *Simulation in Field Testing.* Santa Monica, Rand Corporation.

PRITSKÉR, A. A. B. and KÍVIAT, P. J. (1964) *SIMSCRIPT: A Simulation Programming Language.* Santa Monica, Rand Corporation.

PRITSKÉR, A. A. B. and KÍVIAT, P. J. (1964) *Simulation of Decision Making in Crises.* Santa Monica, Rand Corporation.

PRITSKÉR, A. A. B. and KÍVIAT, P. J. (1965) *The Army Deployment Simulator.* Santa Monica, Rand Corporation.

PRITSKÉR, A. A. B. and KÍVIAT, P. J. (1965) *Simulation and Evaluation of Logistics Systems.* Santa Monica, Rand Corporation.

PRITSKÉR, A. A. B. and KÍVIAT, P. J. (1967) *Digital Computer Simulation.* Santa Monica, Rand Corporation.

PRITSKÉR, A. A. B. and KÍVIAT, P. J. (1967) *Digital Computer Simulation: Statistical Considerations.* Santa Monica, Rand Corporation.

PRITSKÉR, A. A. B. and KÍVIAT, P. J. (1968) *Digital Computer Simulation: Input-Output Analysis.* Santa Monica, Rand Corporation.

PRITSKÉR, A. A. B. and KÍVIAT, P. J. (1969) *Digital Computer Simulation: Estimating Sample Size.* Santa Monica, Rand Corporation.

PRITSKÉR, A. A. B. and KÍVIAT, P. J. (1970) *ECSS: An Extendable Computer System Simulator.* Santa Monica, Rand Corporation.

RAPOPORT, A. (1966) *Two-Person Game Theory: The Essential Ideas.* Lansing, University of Michigan Press.

RAPOPORT, A. (1965) *Fights, Games and Debates.* Lansing, University of Michigan Press.

RAPOPORT, A. (1964) *Strategy and Conscience.* New York, Harper and Row.

RAPOPORT, A. and CHAMMAH, A. M. (1965) *Prisoners Dilemma.* Lansing, University of Michigan Press.

RASER, J. R. (1969) *Simulation and Society: An Exploration of Scientific Gaming.* New York, Allyn & Bacon.

REES, R. W. (1964) *A Statistical Analysis of the Use of a Business Game.* Texas, Texas Technological College.

RICHARDS, L. W. (1970) *Old British Model Soldiers, 1893–1918: An Illustrated Reference Guide for Collectors.* London, Arms and Armour Press.

ROBINSON, J. and BARNES, N. (Ed.) (1967) *New Media and Methods in Industrial Training.* London, B.B.C.

ROCHBERG, R. (1967) *STROP: Player's Manual for JOSS Version.* Santa Monica, Rand Corporation.

ROSSI, P. H. and BIDDLE, B. J. (Ed.) (1966) *The New Media in Education.* Chicago, Aldine Publishing Co.

ROSSI, P. H. and BIDDLE, B. J. (1968) *SIMPAE: A Fast-Time Simulation Package Based on ALGOL.* Farnborough, Royal Aircraft Establishment.

SCHAFFER, M. B. (1966) *Lanchester Models for Phase II Insurgency.* Santa Monica, Rand Corporation.

SCHELLING, T. C. (1960) *The Strategy of Conflict.* Cambridge, Mass., Harvard University Press.

SHAFTEL, F. R. and SHAFTEL, G. (1967) *Role Playing for Social Values: Decision-Making in the Social Studies.* Englewood Cliffs, Prentice-Hall.

SHALOCK, H. D. *et al.* (1964) *Motion Pictures as Test Stimuli: An Application of the New Media to the Prediction of Complex Behaviour.* Monmouth, Oregon State System of Higher Education.

SHEPHARD, R. W. (1966) *Future Trends in War Gaming.* East Coast War Games Council.

SHEPHARD, R. W. (1968) *The Possibilities of Using War Games to Train Army Commanders.*

SHEPHARD, R. W. (1964) *Some Reasons for Present Limitations to the Application of the Theory of Games to Army Problems.*

SHEPHARD, R. W. (1964) *War Games and Simulations.*

SHIGLEY, J. E. (1967) *Simulation of Mechanical Systems: An Introduction.* New York, McGraw-Hill.

SHUBIK, M. (Ed.) (1964) *Game Theory and Related Approaches to Social Behaviour.* New York, Wiley.

SHUBIK, M. (Ed.) (1954) *Readings in Game Theory and Political Behaviour.* New York, Doubleday.

SHUBIK, M. (1959) *Strategy and Market Structure.* New York, Wiley.

SHUCKMAN, A. (1963) *Scientific Decision Making in Business.* New York, Holt, Rinehart & Winston.

SIEGAL, S. and FOURAKER, L. E. (1960) *Bargaining and Group Decision Making: Experiments in Bilateral Monopoly.* New York, Macmillan.

SIMON, H. E. (1966) *Simulations and Gaming of International Military Political Behaviours: Survey of Activities.* Evanston, Northwestern University.

SIMON, H. E. (1966) *TEMPER as a Model of International Relations: An Evaluation of the Joint War Games Agency of the (U.S.) Department of Defence.* Simulatics Corporation.

SISKA, C. P. *et al.* (1954) *Analytic Formulation of a Theater Air Ground Warfare System.* Santa Monica, Rand Corporation.

SMITH, J. (1968) *Computer Simulation Models.* London, Griffin.

SMITH, G. A. and COLE, J. P. (1967) *Bulletin of Quantitative Data for Geographers No. 7: Geographic Games.* Nottingham, Nottingham University.

SMITH, W. N., ESTEY, E. E. and VINES, E. F. (1968) *Integrated Simulation.* London, Edward Arnold.

SMODE, A. F. *et al.* (1963) *Human Factors Technology in the Design of Simulators for Operator Training.* Stanford, Dunlop Associates.

SPECHT, R. D. (1957) *War Games.* Santa Monica, Rand Corporation.

SPENCER, D. D. (1968) *Game Playing with Computers.* New York, Spartan Books.

SPENCER, H. (1887) *Essays on the War Game.*

SNYDER, R. and ROBINSON, J. (1964) *A Comparison of Simulation, Case Studies and Problem Papers in Teaching Decision Making.* Washington, D.C., U.S. Dept of Health, Education & Welfare.

SNYDER, R. *et al.* (1962) *Foreign Policy Decision Making.* New York, Free Press of Glencoe.

STAFFORD, H. A. (1966) *Manufacturing.* Boulder, High School Geography Project.

STOLL, C. S. (1968) *Player Characteristics and Strategy in a Parent-Child Simulation Game.* Baltimore, John Hopkins University.

STRAUCH, R. (1965) *A Preliminary Treatment of Mobile SLBM Defence: A Game Theoretic Analysis.* Santa Monica, Rand Corporation.

STRAUCH, R. (1968) *Pacifist Bargaining Tactics: A Laboratory Assessment of Some Outsider Influences.* Santa Monica, System Development Corporation.

TALACKO, J. (1965) *Introduction to Linear Programming and Games of Strategy.*

TALLEY, H. J. (1967) *A Simulation of Squadron Operations.* Washington, D.C. United States Dept of the Air Force.

TANSEY, P. J. (1970) *Educational Aspects of Simulation.* Maidenhead, McGraw-Hill.

TANSEY, P. J. and UNWIN, D. J. (1969) *Simulation and Gaming in Education.* London, Methuen.

TAYLOR, A. *Discovering War Game Rules.* Aylesbury, Shire Publications.

TAYLOR, A. (1971) *Rules for Wargaming.*

TAYLOR, A. (1970) *Discovering Model Soldiers.*

TAYLOR, A. (1959) *Air Battle Model 2.* Technical Operations Inc.

TAYLOR, A. (1967) *Scenario–Game Model for the Exercise and Evaluation of National Level Civil Defence Systems.* Technical Operations Inc.

TAYLOR, J. L. (1968) *Observations on the Use of Gaming-Simulation Procedures in the Study of Social Systems.* London, London University.

TAYLOR, J. L. (1967) *A Simulation Experiment: Notes on a Prototype Land-Use Gaming Simulation.* Manchester, Manchester University.

TAYLOR, J. L. (1971) *Instructional Planning Systems: A Gaming-Simulation Approach to Urban Problems.* London, Cambridge University Press.

TAYLOR, J. L. and CARTER, K. R. (1969) *Some Instructional Dimensions of Urban Gaming Simulation.* Newcastle-Upon-Tyne, University of Newcastle-Upon-Tyne.

THORELLI, H. B. and GRAVES, R. L. (1964) *International Operations Simulation.* New York, Free Press.

TOCHER, K. D. (1963) *The Art of Simulation.* London, London University Press

TOTTEN, C. A. L. (1880) *Strategos: A Series of American Games Based on Military Principles.*

TROTHA, J. VON (1872) *Introduction to the Employment of the Kriegspiel Apparatus.*

TUNSTILL, J. (1969) *Discovering Wargames.* Aylesbury, Shire Publications.

TWELKER, P. A. (1968) *Interaction Analysis and Classroom Simulation as Adjunct Instruction in Higher Education.* Monmouth, Oregon State System of Higher Education.

TWELKER, P. A. (1968) *Simulation: What is it? Why is it?* San Diego, Association for Supervision and Curriculum Development.

TWELKER, P. A. *et al.* (1967) *Of Men and Machines: Supplementary Guide.* Monmouth, Oregon State System of Higher Education.

TWELKER, P. A. (1963) *Fire Control Simulator.* Washington, D.C., U.S. Dept of Agriculture.

TWELKER, P. A. (1968) *Research and Development of Parallel Hybrid (Computer Simulation) Models.* Washington, D.C., U.S. Dept of the Air Force.

TWELKER, P. A. (1962) *Development of Centuar. A Computerized War Game. Part 1, General Considerations.* Washington, D.C., U.S. Dept of the Army.

TWELKER, P. A. (1967) *Problem-Solving by Digital: Analog Simulation.* Washington, D.C., U.S. Dept of the Army.

TWELKER, P. A. (1967) *The Development of Simulation Materials for Research and Training in Administration of Special Education.* Washington, D.C., U.S. Office of Education.

TWELKER, P. A. (1967) *Instructional Materials.* Columbus, University Council for Educational Administration.

TWELKER, P. A. (1966) *The Jefferson Township School District Simulation.* Columbus, University Council for Educational Administration.

TWELKER, P. A. (1960) *Simulation in Administrative Training.* Columbus, University Council for Educational Administration.

TWELKER, P. A. (1967) *Simulation Techniques in the Evaluation of Clinical Judgement.* University of Illinois.

TWELKER, P. A. (1967) *Materials for the Evaluation of Performance in Medicine.* University of Illinois.

UNWIN, D. (1969) *Media and Methods: Instructional Technology in Higher Education.* Maidenhead, McGraw-Hill.

UNWIN, D. (1968) *The Computer in Education.* London, U.K. Library Association.

UTSEY, J. (1965) *Simulation in Reading: A Breakthrough with Preservice and In-Service Teacher Education.* Eugene, University of Oregon.

VAJDA, S. (1956) *The Theory of Games and Linear Programming.* London, Chapman and Hall.

VANCE, S. (1960) *Management Decision Simulation.* New York, McGraw-Hill.

VENTTSEL, E. S. (1963) *An Introduction to the Theory of Games.*

VERDY DU VERNOIS, J. VON (1899) *The Tactical War Game.*

VETERANS ADMINISTRATION (1961) *The In-Basket Exercise.* Washington, D.C., Veterans Administration.

VOOSEN, B. J. (1965) *Simulation and Evaluation of Logistics Systems.* Santa Monica, Rand Corporation.

VOOSEN, B. J. and CORONA, D. (1949) *Misslogs: A Game of Missile Logistics.* Santa Monica, Rand Corporation.

WALFORD, R. and TAYLOR, J. L. (1972) *Simulation and Gaming in the Classroom.* Harmondsworth, Penguin.

WALFORD, R. (1969) *Games in Geography.* Harlow, Longmans.

WALLEN, C. J. (1966) *Developing Referential Categories with Instructional Simulation.* Monmouth, Oregon State System of Higher Education.

WALLEN, C. J. (1968) *Low Cost Instructional Simulation Materials for Teacher Education.* Monmouth, Oregon State System of Higher Education.

WAR OFFICE (1883) *War Games on Models.* London, War Office.

WEINER, M. G. (1960) *Gaming Limited War.* Santa Monica, Rand Corporation.

WEINER, M. G. (1959) *An Introduction to War Games.* Santa Monica, Rand Corporation

WEINER, M. G. (1969) *Trends in Military War Games.* Santa Monica, Rand Corporation

WEINER, M. G. (1961) *The Use of War Games in Command and Control Analysis.* Santa Monica, Rand Corporation.

WEINER, M. G. (1959) *War Gaming Methodology.* Santa Monica, Rand Corporation.

WELLS, H. G. (1966) *Floor Games: A Companion Volume to 'Little Wars'.*

WELLS, H. G. (1913) *Little Wars.* London, Arms and Armour Press.

WESTERN BEHAVIOURAL SCIENCES INSTITUTE (1966) *An Inventory of Hunches about Simulations as Educational Tools.* La Jolla, Western Behavioural Sciences Institute.

WILKINSON, S. (1887) *Essays on the War Game.*

WILLIAMS, J. D. (1954) *The Complete Strategyst: Being a Primer on the Theory of Games and Strategy.* New York, McGraw-Hill.

WILLINGHAM, J. J. and MALCOLM, R. E. (1965) *Accounting in Action: A Simulation.* New York, McGraw-Hill.

WILSON, A. (1970) *War Gaming.* Harmondsworth, Penguin.

WILSON, A. (1968) *The Bomb and the Computer.* Barry & Radcliff.

WING, R. L. (1966) *Use of Technical Media for Simulating Environments to Provide Individualised Instruction.* Westchester, Board of Co-operative Educational Services.

WISE, T. (1969) *Introduction to Battle Gaming.* Hemel Hempstead, Model and Allied Publications.

YOUNG, J. P. (1956) *A Brief History of War Gaming.* Baltimore, John Hopkins University

YOUNG, J. P. (1959) *A Survey of Historical Developments in War Games.* Baltimore, John Hopkins University.

YOUNG, P. and LAWFORD, P. P. (1967) *Charge! Or How to Play War Games.* London, International Textbooks.

ZOLL, A. A. (1969) *Dynamic Management Education.* New York, Wiley.

3 Synthesis

3.1 The making of a game

'Let there be no mistake about it–a game there must be. Human
and material resources have to be deployed according to some set
of rules.'

Nigel Calder

'It can also be observed that with two circumspect men, one will
achieve his end, the other not; and likewise two men succeed equally
well with different methods . . . '

Niccolò Machiavelli

The model which follows is based on the literature in the field (Clayton and Rosenbloom, 1968; Loveluck, 1969; Glazier, 1969; Walford, 1969; McLeish, 1970; Tansey and Unwin, 1969; Abt, 1970). It shows the stages of production and the relationships between them.

Game production is a craft. The stages represented in the model are the gamer's tools. Careful selection of the appropriate tools and their skilled use leads to a polished end-product. Polish alone is not enough, however; functional design is of prime importance. The needs of functionalism and the constraints of reality may mean accepting a less than perfect finish.

Great games occur when the producer manages to combine inspired design, functional development and a high quality presentation. When these three meet craft is transformed into art.

3.1.1 Aims

Aims consist of broad statements of educational intent. These statements express goals which outline the concepts, structures, processes, facts and attitudes of which it is desired to give the participants experience. The concept 'corridor' is explored in geopolitical terms in the game *Northeast Corridor* produced by Abt Associates Inc. (Taylor, 1970). The *Esso Students Business Game* is an exemplar of those games which deal with structures, in this case of manufacturing and marketing companies (Hargreaves, 1971). *SAM (Simulated Arithmetic Machine)* is intended

MODEL OF THE STAGES IN THE PRODUCTION OF
A GAME OR SIMULATION

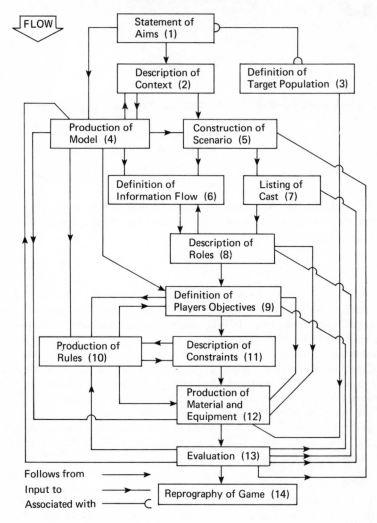

to give insight into the processes involved in the operation of a digital computer (Atherton, 1971). The intention of *Colour* is to teach pupils the facts about composing colours by mixing and about primary and secondary colours (Abt, 1970).

An exercise designed to improve attitudes and make more effective collaboration between experts in different fields is *Professional Collaboration* (Cooke, 1970).

3.1.2 Context

Walford says that 'the prime need at the start is to see a game situation . . ' (Walford, 1972). Clark Abt uses very similar terms (Abt, 1970). Both are using 'see' in the sense of intuitive visualization of the problem involved and its solution. For example, it is quite easy to see the game possibilities in training in communication skills, as in *Radio Covingham* (Jones, 1972). Selection interviewing is another situation in which the gaming possibilities are obvious, as is exemplified in *An Aspect of Choice* (Barber *et al.*, 1969).

Although imagination is essential at this stage, inspiration can be aided by practice. If the teacher is on the 'look-out' for a game situation, he is more likely to recognize the focal point which can be expanded into a 'good' game. The adage 'practice makes perfect' contains a considerable element of truth in the gaming context.

3.1.3 Target population

Whenever educational materials are being produced one of the prime considerations must be those at whom such materials are aimed. Games have to be designed to meet the needs of those being educated, whether or not they can recognize such needs. The report *Half Our Future* (Newsom *et al.*, 1963) stressed the need for education to display immediate relevance. Others have pleaded for a longer-term consideration in defining goals (Tansey, 1971). The beauty of a well-chosen game is that it will meet the need for '. . . teaching for the long range future . . . ' to take place whilst at the same time ensuring that immediate relevance is obtained because '. . . the players learn from their very participation in the game' (Tansey and Unwin, 1969; Boocock and Schild, 1968). In addition, and perhaps most importantly, games motivate the students. In the marketing world it is well recognized that no matter how good a product is, sales will be increased by attractive packaging. Games make ideal packaging for many educational products, with a resultant success in the rate of overcoming student 'sales resistance'. A number of workers have attempted to isolate the reasons for this, and in some cases to quantify the growth of empathy between teacher and taught (Sprague and Shirts, 1966; Garvey, 1971).

3.1.4 Models

The parameters of a game have to be set and this is achieved by the creation of a model. Such a model specifies the variables and indicates the relationships between them. In a sense a model is a formal representation of a theory (George, 1970). These models aid communication, permit investigation, and facilitate prediction (Hanika, 1965; Neil, 1970). Models may be iconic, analogue or symbolic (Hanika, 1965;

Chorley and Haggett, 1967; Hodge, 1970). Games are based on the model selected at this stage of production; sometimes a series of models is employed, and these models may be of varying kinds (Shephard, 1970).

A game based on an iconic model is *Spring Green Motorway* (Finlay, 1972), using the interpretation of iconic in the way in which it is applied at the Administrative Staff College (Hanika, 1965; Walford, 1969). Analogue models are frequently used as the foundation of very effective games, as in *Starpower* (Shirts, 1969). Symbolic models most usually take the form of mathematical expression. Games like *Exercise Lemming* are typical of this approach (Rints and Hayes, 1968).

Whenever models are employed—in other words when producing any game—it is wise to take heed of warnings which have been issued from time to time. 'In an analogue, we self-consciously design into the model a structure that, based on analysis and experimentation, we believe to correspond to some acceptable degree with the real one. . . . In an analogy, we know only that two situations have certain properties in common: we know nothing about the correspondence of the structure of the two situations. . . .. We can usually draw only very weak inferences (if any) by analogy. The inferences being drawn by analogy, however, are sometimes given a degree of credibility they would deserve only if they had been derived from manipulation of an adequate analogue or other type of model' (Ackoff, 1962).

3.1.5 Scenario

Producing the scenario sets the tone of the game within the parameters of the model. Each variable of the model is elaborated sufficiently for a clear picture of the game setting. The scenario may be extremely brief and to the point, as is the case with most mathematics or logic games, for example *On Sets* (Allen, 1969b). On the other hand, the scenario may be extensive, as in *Triangle Trade* (Durham and Crawford, 1969). Short or long, the scenario should be carefully written as it is the point of departure for the game. It is from this material that the participants begin to simulate their roles (Egan *et al.*, 1971).

3.1.6 Information flow

The information processing that constitutes part of any game has to be defined in terms of flow. A flow network has to be described (Katz and Kahn, 1971). Both the amount and the kind of information to be handled have to be specified. It is important that the timing of information presentation be decided—for example, what is to be given initially and what phased throughout the course of play (Love-luck, 1969). The decision as to whether to provide information automatically, on demand or at some cost—in terms of the game—must be made. In some games the information flow networks are made explicit, as in *The Joe Bailey Action Maze* (Zoll, 1969), or they may be implicit, as evidenced in the *Industrial Location Game* (C.S.V., 1972).

3.1.7 Cast list

Games need the contributions of differing interests and groups. The need for these characters will be decided on the basis of the complexity of the model and the possibilities of organizing, structuring and subsequent task delegation.

3.1.8 Role elaboration

Roles are assigned to participants and in order to play out their part they need information about the role. The information can either be presented as part of the preparations, as in a game like *Power Politics* (Durham and Durham, 1970), in which the roles are given in considerable detail. On the other hand, information given in a game like *Mockup* (Bath Youth and Community Service) is very scanty and the participant is required to research his part. Role play involves practising life in the context of the scenario (McPhail, 1972). Role elaboration may involve only an individual part, as in the *Battle of Britain* (Collier, 1969), or the expansion of group positions, as in the *Dynamic Business Game* (Mathewson, 1972) played at the Imperial College of Science and Technology.

3.1.9 Player objectives

Players' objectives should be stated in clear-cut terms which are operational in nature and which have observable outcomes. These goals should state intent and criteria of achievement (Wheeler, 1967). As has been said elsewhere, 'If you're not sure where you're going, you're liable to end up someplace else' (Mager, 1968). The objectives of the players ought to be stated very specifically, as in *Insite II* (Cyros and Langell, 1970). Such objectives may be stated as ultimate goals (Wheeler, 1967), mediate objectives (Bloom *et al.*, 1956) or proximate objectives (Ansoff, 1965).

3.1.10 Statement of rules

The rules are designed to provide a framework which ensures a flow in the operation of the game. The rules are the basis on which players decide what information to transmit and receive, what resources to employ and what actions to take (Glazier, 1969). A list of game rules describes win criteria and any penalties involved. At this stage the rules drawn up are not necessarily those given to the players (Walford, 1972). The final version of rules given to the players will be arrived at after trials to ensure that they work and are understandable. The final list of rules may be long, as in *Equations* (Allen, 1969a), or short as in *The Circuit Game* (Bloomer, 1971), but they must be absolutely clear and unambiguous.

3.1.11 Description of constraints

Those factors which affect or stop the players objectives being achieved are described—for example, the use of chance cards, as in *Monopoly* (Darrow, 1935), or by restricted play on a time period basis, as in *Inlogov Local Authority Game* (Armstrong, 1970). A game which includes all the factors present in reality is very difficult to play; hence constraints are introduced to restrict those variables that make play difficult. The constraints should be introduced in such a way as to limit the abstraction of the game from reality as much as possible.

3.1.12   Materials and equipment

Items of equipment (e.g. dice, boards, etc.) and materials (e.g. books, maps, etc.) are combined in such a way as to make the presentation of the game package as suitable for its purpose and as enjoyable to use as possible.

3.1.13   Evaluation

Evaluation is an iterative process designed to assess a number of variables. To begin with, criteria for each variable are determined and an acceptable measure for each is decided on. Additionally, the size of each sample of the student population should be predetermined.

The process of evaluation enables an assessment to be made of the amount of learning, its quality and efficiency, student interest, enjoyment and involvement, ease of organization and play, content validity, reliability and acceptability by the tutor (Gagné, 1965; Eraut, 1970; Davies, 1971; Wheeler, 1967; Rackham *et al.*, 1971; Hartley, 1972).

3.1.14   Reprography

At this stage the game is produced in sufficient numbers for the purposes of its designer, at a standard which will satisfy the needs of those involved.

3.1.15   Bibliography

ABT, C. C. (1970) *Serious Games.* New York, Viking Press.
ACKOFF, R. L. (1962) *Scientific Method: Optimizing Applied Research Decisions.* New York, Wiley.
ALLEN, L. E. (1969a) *Equations.* New Haven, Wff'n Proof Inc.
ALLEN, L. E. (1969b) *On Sets.* New Haven, Wff'n Proof Inc.
ANSOFF, H. I. (1965) *Corporate Strategy.* Harmondsworth, Penguin.
ARMSTRONG, R. H. R. (1970) in ARMSTRONG, R. H. R. and TAYLOR, J. L. (Ed.) *Instructional Simulation Systems in Higher Education.* Cambridge, Cambridge Institute of Education.
ATHERTON, R. T. (1971) in TANSEY, P. J., *Educational Aspects of Simulation.* Maidenhead, McGraw-Hill.
BARBER, D. *et al.* (1969) in TANSEY, P. J. and UNWIN, D., *Simulation and Gaming in Education.* London, Methuen.
BATH YOUTH AND COMMUNITY SERVICE *Mockup.* Bath, City of Bath Youth and Community Service.

BLOOM, B. *et al.* (1956) *Taxonomy of Educational Objectives.* London, Longmans.
BLOOMER, J. (1971) *The Circuit Game,* Glasgow, Glasgow University.
BOOCOCK, S. S. and SCHILD, E. O. (Ed.) (1968) *Simulation Games in Learning.* Beverly Hills, Sage Publications.
COMMUNITY SERVICE VOLUNTEERS (1972) *Industrial Location Game.* London, C.S.V.
CHORLEY, R. J. and HAGGETT, P. (1967) *Models in Geography: The Madingley Lectures for 1965.* London, Methuen.
CLAYTON, M. and ROSENBLOOM, R. (1968) in BOOCOCK, S. S. and SCHILD, E. O. (Ed.) *Simulation Games in Learning.* Beverly Hills, Sage Publications.
COLLIER, B. (1969) *Battle of Britain.* London, Jackdaw Publications Ltd.
COOKE, J. E. (1970) in ARMSTRONG, R. H. R. and TAYLOR, J. L. (Ed.) *Instructional Simulation Systems in Higher Education.* Cambridge, Cambridge Institute of Education.
CYROS, K. L. and LANGELL, J. A. (1970) *Insite II.* Cambridge, Mass., M.I.T.
DARROW, C. W. (1935) *Monopoly.* Salem, Parker Bros. Inc.
DAVIES, I. K. (1971) *The Management of Learning.* Maidenhead, McGraw-Hill.
DURHAM, R. and CRAWFORD, D. J. (1969) *Triangle Trade.* Monmouth, Oregon State System of Higher Education.
DURHAM, R. and DURHAM, V. (1970) *Power Politics.* Monmouth, Oregon State System of Higher Education.
EGAN, J. M. *et al.* (1971) *Simulex II.* Durham, New Hampshire Council of World Affairs.
ERAUT, M. (1970) in TAYLOR, G. (Ed.) *The Teacher as Manager.* London, N.C.E.T.
FINLAY, J. (1972) in LONGLEY, C. (Ed.) *Games and Simulation.* London, B.B.C.
GAGNÉ, R. M. (1965) in GLASER, R. (Ed.) *Training Research and Education.* New York, Wiley.
GARVEY, D. M. (1971) in TANSEY, P. J. (Ed.) *Educational Aspects of Simulation.* Maidenhead, McGraw-Hill.
GEORGE, F. (1970) *Models of Thinking.* London, Allen & Unwin.
GLAZIER, R. (1969) *How to Design Educational Games.* Cambridge, Mass., Abt Associates Inc.
HANIKA, F. de P. (1965) *New Thinking in Management.* London, Lyon, Grant & Green.
HARGREAVES, S. N. (1971) in ARMSTRONG, R. H. R. and TAYLOR, J. L. (Ed.) *Feedback on Instructional Simulation Systems.* Cambridge, Cambridge Institute of Education.
HARTLEY, J. (Ed.) (1972) *Strategies for Programmed Instruction: An Educational Technology.* London, Butterworths.

HODGE, H. P. R. (1970) in BAJPAI, A. and LEEDHAM, J. F. (Ed.) *Aspects of Educational Technology IV*. London, Pitman.

JONES, J. K. (1972) in LONGLEY, C. (Ed.) *Games and Simulation*. London, B.B.C.

KATZ, D. and KAHN, R. L. (1971) in McRAE, T. W. (Ed.) *Management Information Systems*. Harmondsworth, Penguin.

LOVELUCK, C. (1969) *Notes on the Construction, Operation and Evaluation of Management Games*. Weybridge, Management Games Ltd.

McLEISH, J. (1970) in ARMSTRONG, R. H. R. and TAYLOR, J. D. (Ed.) *Instructional Simulation Systems in Higher Education*. Cambridge, Cambridge Institute of Education.

McPHAIL, P. (1972) in LONGLEY, C. *Games and Simulations*. London, B.B.C.

MAGER, R. F. (1968) *Developing Attitude Toward Learning*. Palo Alto, Fearon.

MATHEWSON, S. C. (1972) *Dynamic Business Game*. London, Imperial College of Science and Technology.

NEIL, M. W. (1970) in BAJPAI, A. and LEEDHAM, J. F. (Ed.) *Aspects of Educational Technology IV*. London, Pitman.

NEWSOM, J. H. *et al.* (Ed.) (1963) *Half Our Future*. London, H.M.S.O.

RINTS, N. and HAYES, D. (1968) *Exercise Lemming*. Slough, Slough College of Technology.

RACKHAM, N. *et al.* (1971) *Developing Interactive Skills*. Guilsborough, Wellens Publishing.

SHEPHARD, R. W. (1970) in ARMSTRONG, R. H. R. and TAYLOR, J. L. (Ed.) *Instructional Simulation Systems in Higher Education*. Cambridge, Cambridge Institute of Education.

SHIRTS, R. G. (1969) *Starpower*. New York, Western Behavioural Sciences Institute.

SPRAGUE, H. T. and SHIRTS, R. G. (1966) *Exploring Classroom Uses of Simulation*. New York, Western Behavioural Sciences Institute (mimeo).

TANSEY, P. J. and UNWIN, D. J. (1969) *Simulation and Gaming in Education*. London, Methuen.

TANSEY, P. H. (Ed.) (1971) *Educational Aspects of Simulation*. Maidenhead, McGraw-Hill.

TAYLOR, J. L. (1970) in ARMSTRONG, R. H. R. and TAYLOR, J. L. (Ed.) *Instructional Simulation Systems in Higher Education*. Cambridge, Cambridge Institute of Education.

WALFORD, R. (1969) *Games in Geography*. London, Longmans.

WALFORD, R. (1972) in LONGLEY, C. (Ed.) *Games and Simulations*. London, B.B.C.

WHEELER, D. K. (1967) *Curriculum Process*. London, U.L.P.

ZOLL, A. A. (1969) *Dynamic Management Education*. Reading, Mass., Addison-Wesley.

3.2 Training and sustaining

'If the task of the teacher is often satisfying, it is sometimes frustrating and debilitating, it requires the maintenance of a fairly high emotional temperature for lengthy periods of time, and the conventional timetable provides for only minimal contact with fellow practitioners.

William Taylor

'You do not learn to swim, to play tennis, to drive a motor car or to make love simply by reading books or being told; to do these things, and especially to become competent at them, they have to be practised.'

Terry Morgan

3.2.1 Course organizers

The following details cover a small sample of the organizations and institutions who offer courses in gaming and simulation, or courses of which gaming and simulation form a substantial part.

Abt Associates Inc.
    55 Wheeler Street, Cambridge, Massachusetts 02138, U.S.A.
Academic Games Associates
    430 East 33rd Street, Baltimore, Maryland 21218, U.S.A.
Bath University of Technology
    School of Education, Upper Borough Walls, Bath, Somerset.
Behavioural Sciences Laboratory
    College of Social and Behavioural Sciences, Ohio State University, 404B West 17th Avenue, Columbus, Ohio 43210, U.S.A.
Berkshire College of Education
    Woodlands Avenue, Earley, Reading, Berkshire
British Broadcasting Corporation
    Broadcasting House, Portland Place, London W.1.
British Institute of Management
    Management House, 80 Fetter Lane, London E.C.4.
Coventry Technical College
    Butts, Coventry CV1 3GD.
Dunchurch Industrial Staff College
    Dunchurch, Rugby, Warwickshire
Games Group, Mental Health Research Institute
    University of Michigan, Ann Arbor, Michigan 48104, U.S.A.
Imperial College of Science and Technology
    Dept. Management Science, Exhibition Road, London SW7 2BX.

Information Resources Inc.
    1675 Massachusetts Avenue, Cambridge, Massachusetts 02138, U.S.A.
Instructional Simulations Inc.
    2147 University Avenue, St Paul, Minnesota 55104, U.S.A.
Kaleidoscope
    6 Harley Road, London N.W.3.
Kansas Educational Simulation Center
    Division of Social Sciences, Kansas State Teachers College, Emporia,
    Kansas 66801, U.S.A.
Learning Games Associates
    2253 Medford Road, Ann Arbor, Michigan 48104, U.S.A.
Loughborough University of Technology
    Centre for Extension Studies, Loughborough, Leicestershire.
Leisure Learning, ORD Ltd
    15 Crestview, Dartmouth Park Hill, London NW5 1JB.
Management Games Ltd
    70 Baker Street, Weybridge, Surrey.
Ministry of Defence (Army)
    H.Q. Western Command, Education Division, Chester.
National Academic Games Project
    Nova University, S.W. College Avenue, Fort Lauderdale, Florida 33314,
    U.S.A.
National Gaming Council
    Center for Social Organization of Schools, The John Hopkins University,
    3505 N. Charles Street, Baltimore, Maryland 21218, U.S.A.
Programmed Instruction Centre for Industry
    32 Northumberland Avenue, Sheffield S10 2TX.
Radius: Research Analysis and Design in Urban Systems
    6 Willoughby Road, London N.W.3.
Real World Learning Inc.
    134 Sunnydale Avenue, San Carlos, California 94070, U.S.A.
Sheffield Polytechnic
    Centre for Management Studies, Grove Road, Totley Rise, Sheffield S17 4DJ.
Simulation Systems Program
    Teaching Research Division, Oregon State System of Higher Education,
    Monmouth, Oregon 97361, U.S.A.
Society for Academic Gaming and Simulation in Education and Training
    5 Errington, Moreton-in-Marsh, Gloucestershire.
University of Birmingham
    Institute of Local Government Studies, PO Box 363, Birmingham B15 2TT.
University of Glasgow
    Department of Education, 4 University Gardens, Glasgow W.2.

University of Leicester
School of Education, 21 University Road, Leicester LE1 7RF.
University of Liverpool
Audio-Visual Aids and Programmed Learning Unit, Chatham Street, Liverpool L69 38X.
University of Ulster
Education Centre, Coleraine, Northern Ireland.
University of Utrecht
Department of Sociology, PO Box 13015, Utrecht, Holland.

3.2.2 Associations, clubs and societies

The following details are of a few of the organizations in the field. There must be many other organizations who ought to appear in this register but unfortunately there are no details at hand. It is hoped that these details will be brought to our attention for inclusion in the next edition.

S.A.G.S.E.T.

The Society for Academic Gaming and Simulation in Education and Training
5 Errington, Moreton-in-Marsh, Gloucestershire.

Council

The Society is administered by an elected council and officers.

Chairman		
G. Mallen	System Simulations Ltd.	1970-73
Honorary Secretary		
G. I. Gibbs	Home Office Unit for Educational Methods	1970-73
Honorary Treasurers		
K. Payne	London Schools	1970-71
A. Warrington (Mrs.)	Maria Grey College of Education	1972-73
Council members		
P. J. Tansey	Berkshire College of Education	1970-72
A. Aldrich	Wiltshire Youth Service	1970-71
H. H. G. Nisbet (Maj.)	R.A.E.C.	1970-72
K. Jones	Freelance Journalism	1971-72
M. Powell	Essex Schools	1971-72
D. D. Simmons	Further Education Staff College	1971-72
G. Parsons	Polytechnic of the South Bank	1971-73
D. Teather (Dr.)	Liverpool University	1971-73
P. G. Dean	University of London	1971-73

J. L. Taylor (Dr.)	Sheffield University	1972-73
M. van Ments	Loughborough University	1972-73
R. J. Talbot	Manchester University	1972-73

History

During a gaming course in 1969 at Bulmershe College of Education a group of those attending and the tutors agreed to keep in touch informally. Throughout the next twelve months contact was maintained and two of the founder members of this informal group organised a summer conference, held at Reading.

Whilst this Conference was in progress it became apparent that the informal meeting had been useful and others wished to join in. The increase in members meant that a more organised arrangement was required and so the Society was born.

Aims and activities of the society

The Society exists to encourage and develop the techniques of gaming and simulation in all areas of education and training, including selection and assessment. It does this by promoting communication amongst practitioners through meetings and publications.

The Society also aims to establish strong links between practitioners and those carrying out relevant research in fields related to education. This it does through conferences, the publication of bibliographies of relevant material and the establishment of study groups which review problem areas and produce reports.

Publications

The main organ of the Society is a quarterly newsletter circulated free to all members. This contains descriptions of games and simulations, information about sources and availability of games materials, news about relevant meetings, courses and conferences, news about research work and requests for games materials. A bibliography covering the subject of games and simulations in education and training is published.

Conferences

The conferences aim to communicate the most up to date ideas and techniques concerned with the theory and practice of educational games and simulations. They are designed always to mix theory with practice so the conferences present the opportunity to become acquainted with new games and techniques by participation and by listening to papers from invited experts.

Membership

Membership is open to all interested in the application and development of simulation and gaming techniques in training and education.

Apply to:- The Secretary,
5 Errington,
Moreton-in-Marsh,
Gloucestershire.

The Adult Postal Wargame Club
32 Saxon Close, East Preston, Sussex. (Jones, M.)

American Council on Educational Simulation and Gaming
PO Box 5131, Industrial Station, 453 North Snelling Avenue, St Paul, Minnesota 55104, U.S.A. (Twelker, P.)

American Educational Research Association: Simulation Systems Group
University of Kentucky, Lexington, Kentucky, U.S.A. (Dettre, J. R.)

Avalon Hill Intercontinental Kreigspiel Society
22 Salisbury Road, Seaford, Sussex.

The British Diplomacy Club
PO Box 4, London N6 4DF

British Go Association
60 Wantage Road, Reading, Berkshire.

British Table Soccer Association
30 Saltmarsh Lane, Hayling Island, Hampshire. (Maison, J. D.)

Cambridge University Go Society
St John's College, Cambridge. (Smith, R. J.)

Chase Militaria Society
5 Nailers Drive, Burntwood, Walsall, Staffordshire.

Diplomacy Sub-Society
Intervarsity Club, 117 Queensway, London W.2.

English Table Soccer Association
1 Salford Road, Old Marston, Oxford OX3 ORX. (Winston, B. M.)

Glasgow and District Wargaming Society
2253 London Road, Glasgow G32 8HW. (Johnston, J.)

Harlow Wargamers Group
Tye Green Community Association

London Go Club
Royal Oak, 88 Bishops Bridge Road, London W.2

London Wargamers Section
Ordnance Arms, York Road, London S.E.1.

Manchester Tactical Society
Department of Military Studies, Dover Street, Manchester M13 9PL. (Bateman, E.)

Manchester War Games Society
Department of Military Studies, Dover Street, Manchester M13 9PL.
(Taylor, D.)
National Games Club
Bedford Corner Hotel, Bayley Street, London W.C.1.
National Gaming Council
1100 17th Street N.W., Washington, D.C. 20036, U.S.A.
Naval Wargamers Society
Ordnance Arms, York Road, London S.E.1.
Oxford Wargaming Society
48 Churchill Way, Long Hanborough, Oxfordshire.
Postal Scrabble Association
11 Tottenham Court Road, London W1A 4XF.
Scottish Table Soccer Association
41 Penicuik Road, Roslin, Midlothian. (Spratt, G.)
Sealed Knot Society of Cavaliers
Highcroft, 55 Graham Road, Great Malvern, Worcestershire.
(Lt.-Col. Hastings-Read)
Simulation Councils Inc.
1010 Pearl Street, PO Box 2228, La Jolla, California 92037, U.S.A.
The Society of Ancients
757 Pershore Road, Birmingham B29 7NY. (Barker, P.)
Tunbridge Wells Wargames Society
154d Upper Grosvenor Road, Tunbridge Wells, Kent.
War Games Research Group
75 Ardingley Drive, Goring-By-Sea, Sussex. (O'Brien, B.)
Warwick University War Games Society
Room E61, Rootes Hall, Warwich University, Coventry, Warwickshire.

3.3 Using games and simulations

'Adults sometimes just "play", but children just "play" far more.'
Susanna Millar

'Pastimes and games are substitutes for the real living of real intimacy.
Because of this they may be regarded as preliminary engagements rather
than as unions, which is why they are characterized as poignant forms of
play.'
Eric Berne

'A good simulation system replaces a real-life situation by substituting
elements as closely similar as possible . . . or it replaces the real situation by
a simpler, more easily manipulated system to facilitate study.'
Ruth Beard

What are the functions of simulations? There are three widely accepted uses to which games and simulations can be put. These are teaching, assessment and research.

3.3.1 Teaching

Games and simulations may be employed for a variety of purposes within the teaching-learning context. One basic and very important use is the building of self-confidence (Garvey, 1971). Increasing motivation is also an important purpose of gaming (Davies, 1971). Understanding can be improved by the use of a well-chosen game (Tansey, 1971). Perhaps the use of games and simulations for engendering learning is the most widely accepted purpose of gaming (Boocock, 1968). Finally, games and simulations are a means of giving meaningful practice of that which has been learnt (Varenhorst, 1968).

At the simplest level games may be employed to impart factual information (Lukács and Tarján, 1963). Such a game is *Formulon* (Stoane, 1972). Other games—for example, *Math Magic* (Mazer and Bolton, 1954)—may be used to practice skills (Gagné, 1965). A very important use of games is their employment in the teaching of concepts (Abt, 1970). An example of this usage is *Science Maps* (Gibbs, 1972). Principles may also be taught by simulations and games, as *Coexist* (Dean, 1972) demonstrates admirably (Zoll, 1966). Values may be imparted via the gaming media (McPhail, 1972), as in *The Value Game* (Lineham and Irving, 1970). *Sunshine* (Yount and De Kock, 1969) is a game which illustrates the learning which takes place during its use in changing attitudes (McCormick, 1972).

3.3.2 Evaluation

Evaluation is used here to mean the use of measures of proficiency of various kinds, the diagnosis of problems and the prognosis of future performance.

The assessment of learning is probably the greatest area of usage of games for evaluation (Dimitriou, 1971). A game which has been employed in this way is *Tenement* (Cooper, 1972). Fairly wide use of games for the diagnosis of problems also occurred (McAleese and Unwin, 1971). For instance, *The Five Game* (Springle, 1967) has been used like this. Prognosis is a less widely developed aspect of using games for evaluation, but some work has been done in this field (Wilson, 1970). The game *An Aspect of Choice* (Barker *et al.*, 1969) has been used for this purpose.

3.3.3 Research

Simulations and games may be employed for the purposes of research, either pure or applied. The pure research use of gaming may allow description of variables, their classification and finally their quantification (Gibbs, 1972). Pure research usage of gaming is described in the communications field (Rae, 1970). The game *Simpol* (Mallen, 1970) has been used for pure research into organizational theory.

Research by system simulation has allowed the development of operational rationales (Gagné, 1965). For example, the game *STAGE* (*Simulation of Total Atomic Global Exchange*) (U.S. Dept of Defence, 1962) has been employed to plan strategy. Applied research uses of simulation games allow investigations of the validity of theses (Wilson, 1970). Reliability testing is a further applied research function of gaming (Chapman *et al.*, 1959). Simulations may be used for the diagnosis of problems, which is the way *Saga* (Higson, 1969) has been used. Finally games may be employed to forecast future outcomes (Calder, 1969).

3.3.4 Bibliography

ABT, C. C. (1970) *Serious Games.* New York, Viking Press.
BARBER, D. *et al.* (1969) in TANSEY, P. J. and UNWIN, D. *Simulation and Gaming in Education.* London, Methuen.
BOOCOCK, S. S. (1968) in BOOCOCK, S. S. and SCHILD, E. O. (Ed.) *Simulation Games in Learning.* Beverly Hills, Sage Publications.
CALDER, N. (1969) *The Environment Game.* London, Panther.
CHAPMAN, R. L. *et al.* (1959) *The System Research Laboratory's Air Defence Experiments.* New York, Management Science.
COOPER, G. (1972) *Tenement.* London, Shelter.
DAVIES, I. K. (1971) *The Management of Learning.* Maidenhead, McGraw-Hill.
DEAN, P. G. (1972) *Coexist.* Liverpool, SAGSET News.
DIMITRIOU, B. (1971) in ARMSTRONG, R. H. R. and TAYLOR, J. L. (Ed.) *Feedback on Instructional Simulation Systems.* Cambridge, Cambridge Institute of Education.
GAGNÉ, R. M. (1965) in GLASER, R. (Ed.) *Training Research and Education.* New York, Wiley.
GARVEY, D. M. (1971) in TANSEY, P. J. (Ed.) *Educational Aspects of Simulation.* Maidenhead, McGraw-Hill.
GIBBS, G. I. (1972) *Notes on Simulation and Gaming.* Sheffield, Programmed Instruction Centre for Industry.
GIBBS, G. I. (1972) *Science Maps.* Moreton-in-Marsh (mimeo).
HIGSON, L. M. (1969) *Saga.* Radlett, Epic Educational Toys.
LINEHAM, T. E. and IRVING, W. S. (1970) *The Value Game.* New York, Herder and Herder.
LUKÁCS, C. and TARJÁN, E. (1970) *Mathematical Games.* London, Pan Books.
McALEESE, W. R. and UNWIN, D. (1971) in PACKHAM, D. *et al.* (Ed.) *Aspects of Educational Technology V.* London, Pitman.
McCORMICK, J. (1972, July) *Simulation and Gaming as a Teaching Method.* London Programmed Learning and Educational Technology.
McPHAIL, P. (1972) in LONGLEY, C. (Ed.) *Games and Simulations,* London, B.B.C.
MALLEN, G. L. (1970) *Simpol.* Richmond, Systems Research Ltd.

MAZER, E. E. and BOLTON, D. R. (1954) *Math Magic*. Chicago, Cadaco Inc.

RAE, J. (1970) in ARMSTRONG, R. H. R. and TAYLOR, J. L. (Ed.) *Instructional Simulation Systems in Higher Education*. Cambridge, Cambridge Institute of Education.

SPRINGLE, H. A. (1967) *The Five Game*. Chicago, Science Research Associates.

STOANE, J. S. (1972) *Formulon*. Letham, Chemical Teaching Aids.

TANSEY, P. J. (Ed.) (1971) *Educational Aspects of Simulation*. Maidenhead, McGraw-Hill.

U.S. DEPT. OF DEFENCE. (1962) *STAGE (Simulation of Total Atomic Global Exchange)*. Washington, D.C., U.S. Dept of Defence.

VARENHORST, B. B. (1968) in BOOCOCK, S. S. and SCHILD, E. O. (Ed.) *Simulation Games in Learning*. Beverly Hills, Sage Publications.

WILSON, A. (1970) *War Gaming*. Harmondsworth, Penguin.

YOUNT, D. and DE KOCK, P. (1969) *Sunshine*. Lakeside, Interact.

ZOLL, A. A. (1969) *Dynamic Management Education*. Reading, Mass., Addison-Wesley.

4 Prognosis

4.1 Problems and solutions

'The most important element in using dynamic education techniques successfully is an experimental attitude.'

A. A. Zoll

Two major problems seem to remain fairly intractable. Firstly, does transfer from the game/simulation to real life take place? Secondly, can effective evaluation techniques be applied to games and simulations?

Had this been written a few years ago, one could have said that it was not possible to answer the question about transfer. Even now one might be justified in having reservations. As yet there is little empirical evidence to support the view that transfer does take place. There is, however, promising work in this field and the work of Gaffya, Cruickshank and Broadbent provides evidence of positive transfer (Cruickshank, 1971).

A more intractable problem is that of evaluation. Firstly, one must specify what kind of evaluation one is talking about. An evaluation may be concerned with the acceptability of a game or simulation to the student. Again, it may be concerned with motivation. Finally, evaluation may be concerned with the quality and/or quantity of learning.

The acceptability to the student of games and simulations has been reported on many occasions but concrete measurement is rather less evident. However, Cherryholmes has shown that this learning method is accepted and enjoyed by bright students and Coleman has shown similar results with less able students. It would seem that games and simulations are learning media which most people find acceptable.

The most widely reported advantage of games and simulations, and probably the one with the least hard evidence to support its claim, is their motivating power. Perhaps all one can do is ask 'How do you measure motivation?' Perhaps the work of Sprague and Shirts indicates the way (Garvey, 1971).

Finally, we come to examine the problem of evaluating the quality and quantity of learning. A certain amount of work has been done in evaluating the quality and quantity of learning in situations which are deterministic (Gibbs, 1972) and a little has been attempted in the measurement of learning using simulations of which the outcomes are probabilistic (Gibbs, 1972). Perhaps the most promising work so far done has been on the measurement of the kinds and amounts of interaction between the participants of a simulation using the techniques developed at British European Airways Ltd and at International Computers Ltd (Rackham *et al.*, 1971).

Despite the efforts so far made in this area of measurement it is apparent that there is only limited empirical evidence available. The duration of experiments has been short, both in terms of the playing time of the game/simulation used and in terms of the total time span over which testing has been carried out. Usually testing has been based on the use of only one game/simulation or at the most two. Additionally in almost every case the size and nature of the test population has been too limited to draw conclusions which are generally valid.

What, then, are the needs of the gaming/simulation movement in the foreseeable future? A considerable amount of research is required into effective ways of evaluating the benefits of games and simulations. The development of many more games and simulations is required. These must extend the content range, the range of students they are aimed at and the variety of forms. Finally, it is necessary that a knowledge of work in the field is disseminated as widely as possible. This book is intended to meet this last need, at least in part.

4.2 Bibliography

CRUICKSHANK, D. R. (1971) in TANSEY, P. J. (Ed.) *Educational Aspects of Simulation.* Maidenhead, McGraw-Hill.

GARVEY, D. M. (1971) in TANSEY, P. J. (Ed.) *Educational Aspects of Simulation.* Maidenhead, McGraw-Hill.

GIBBS, G. I. (1972a) *Simulation Exercises in the Learning of Legislation.* Moreton-in-Marsh (mimeo).

GIBBS, G. I. (1972b) *Assessment in Breathing Apparatus Training.* Moreton-in-Marsh (mimeo).

RACKHAM, N. *et al.* (1971) *Developing Interactive Skills.* Guilsborough, Wellens Publishing.

5 Register of Games and Simulations

5.1 Games and simulations

A1. THE ABRAHAM LINCOLN ELEMENTARY PRINCIPAL-SHIP; USA
–, University Council for Educational Administration
Educational administrators

A2. ABSTRACTION; UK
F. C. Davis; Albion
General
7
Based on 'Diplomacy'; a geographical rationalisation

A3. ACCOUNTING IN ACTION: A SIMULATION; 1965; USA
J. J. Willingham, R. E. Malcolm; McGraw-Hill
College, management

A4. ACHTHERSNO VILLAGE; UK
–; Scottish Youth and Community Service Information Centre
Youth leaders
Running a youth club

A5. ACQUIRE; USA
S. Sackson; 3M Company
General
2-6; 1¼ hours
Competitive property investment and acquisition

A6. ACTION; 1971; UK
C. Grant; A. & C. Black
Military
2
Practice of and insight into tactics of the Seven Years War

A7. THE ACTION CORPORATION; 1971; USA
W. Kell, A. McCosh; Ronald Press Co
College students
12+; 8-24 hours
Integration of accounting knowledge and business understanding

A8. ACTION FOR LIBEL; 1971; UK
J. K. Jones; J. K. Jones
Secondary school
Insight into law and roles in libel cases

A9. AD GAME; 1970; USA
M. Hanan; Hanan & Son
Management
1+; 30-60 minutes
Practice in advertising planning and performance

A10. AD-LIB CROSSWORD CUBES; UK
–; Waddingtons Ltd
General
1+
Teaches spelling and vocabulary

A11. AD PLAN; 1960; USA
M. Hanan; Hanan & Son
Management
15-45; 7 hours
Training in management knowledge and skills in advertising management

A12. ADMINSTRATORSHIP; USA
–; Solidarity House Inc
College, management
Administration skills

A13.   ADMIRALS; USA
−; Parker Bros
General
Naval tactics

A14.   ADVENTURING; USA
−; Abt Associates Inc
Junior high school
Insight into structure of English society
    prior to the Civil War

A15.   ADVENTUROUS EQUATIONS;
          1969; USA
L. A. Allen; Wff'n Proof Inc
Grades 4-12
2+
Practice in addition, subtraction,
    multiplication, division, exponentiation
    and radicals and the use of different
    number bases

A16.   AERO SPACE BUSINESS
          ENVIRONMENT SIMULATOR
          (ABES); USA
R. K. Summit; Lockheed Palo Alto
    Research Laboratory; IBM
Management, military
5-50; 150 minutes+
Practice in working and planning using
    objectives and policies

A17.   AFTER YOU'VE GONE, USA
−; Visiting Nurse Service, N.Y.
Trainee nurses
Insight into the problems of a mother
    deciding whether or not to abort a
    child

A18.   A.F.W.E.T. (AIR FORCE
          WEAPONS EFFECTIVENESS
          TESTING); 1967; USA
−; Raytheon
Air force
Practice in air combat tactics

A19.   AFRIKA KORPS; 1964; USA
−; Avalon-Hill Inc
General
2+; 2-4 hours
Illustrates the relationship between tactics
    and logistics in the context of the North
    African Campaigns of Rommel

A20.   AFRIKA KORPS II; USA
C. Lane; Panzerfaust
General
Logistics of 1940-1942 campaigns

A21.   AGILE-COIN; 1964; USA
−, Abt Associates Inc
Military-political
Insight into counter-insurgency campaigns
    in South East Asia

A22.   A1 PRACTICE; 1962, USA
L. E. Allen; Wff'n Proof Inc
1st grade to college
2+
Teaches use of disjunction-in rule of logic

A23.   THE AID COMMITTEE
          GAME: BOTSWANA; 1970; UK
O. G. Thomas, M. F. C. Clarke; Oxfam
Secondary school, students, adults
20
Insight into the processes of deciding
    priorities in aid-giving

A24.   THE AID COMMITTEE
          GAME: UPPER VOLTA;
          1970; UK
O. G. Thomas, M. F. C. Clarke; Oxfam
Secondary school, students, adults
20
Insight into the process of deciding
    priorities in aid-giving

A25.   AIR CANADA MANAGEMENT
          GAME; 1964; Canada
A. A. Lackman, H. J. G. Whitton; Air
    Canada
Management
10-20; 1-4 days
Insight into relationship of departmental
    functions, scientific tools of manage-
    ment, techniques and the newer theories

A26.   AIR CHARTER; UK
−; Waddingtons Ltd
8 years old and over
2-4
Players compete to corner the air freight
    market in Australasia and S.E. Asia

A27.   AIRPORT; USA
—; Dynamic Design Industries
General
Running an airport

A28.   AIRPORT CONTROVERSY;
1971; UK
J. K. Jones; J. K. Jones
Secondary school
Insight into public planning problems, pressure
groups and public inquiries

A29.   ALEA; 1969; UK
R. H. R. Armstrong, M. Hobson; Institute
of Local Government Studies, Birmingham
University
Local government officers
30+; 10-12 hours
Study of new city development and its
impact on the planning and administrative
problems of an existing town confronted
with the problems of urban renewal

A30.   ALESIA; 1970; USA
R. Bradley; R. Faubert, Cambridge
Roman conquest of Gaul

A31.   ALEXANDER THE GREAT;
USA
—; Guidon Games
General
War game of the campaigns of Alexander

A32.   ALL FALL DOWN; UK
—; Waddingtons Ltd
5 years old and over
2-4
Teaches tactical thinking

A33.   ALPHABET PLAYTRAY; UK
—; Victory Ltd
3-6 years
1
Practice in matching shapes

A34.   THE AMERICAN
CONSTITUTIONAL
CONVENTION; 1969; USA
L. Stitelman, W. D. Coplin; Science
Research Associates
High school, college
32-49; 2-6 hours
Insight into political problems facing the
Founding Fathers and into the political
philosophy which influences any political
system and constitution making

A35.   AMERICAN GOVERNMENT
SIMULATIONS; USA
W. D. Coplin, L. Stitelman; Department of
Political Science, Syracuse University;
Science Research Associates
College
30+
Insight into history and present day nature
of US government

A36.   AMERICAN MANAGEMENT
ASSOCIATION GAME; USA
—; American Management Association
Management

A37.   AMERICAN MANAGEMENT
ASSOCIATION GENERAL
MANAGEMENT BUSINESS
SIMULATION (THE MOSE
COMPANY); 1957; USA
—; American Management Association
Management
5-168; 3-20 hours
Decision management and group dynamics

A38.   AMERICAN REVOLUTION; USA
—; Simulations Publications
General
Gives insight into the American war of
independance

A39.   AMERICARD—USA; 1968; USA
R. Allen; Wff'n Proof Inc; International
Learning Corporation
Primary, junior high and high school
2-4; 1-5 hours
Learning to find and use information about
the size, shape and location of the US
states and their geography, history,
constitutions and current affairs

A40.   AMOEBA, UK
–; Educational Supply Association
Primary school and upwards
Shape recognition

A41.   ANAGOG; 1972; Denmark
Piet Hein; Piet Hein
General
1
Six pieces of joined unit spheres are to be
    formed into a 20 sphere tetrahedron, two
    10 sphere tetrahedrons or other figures

A42.   . . . AND GLADLY TEACH;
        UK
W. Taylor, S. Moore; Harlech T.V.;
    EVK Partnerships
Teachers
Insights into the sociology of the classroom

A43.   ANDESIA; USA
J. Reese; J. Reese
5th and 6th grade
18-30; 4-6 hours
Insight into the problems and conflicts of a
    middle-Andes nation

A44.   ANDLINGER BUSINESS
        GAME; USA
–; Harvard University
University, management
2½ days
Establishing and running a company

A45.   ANIMAL LOTTO; UK
–; Educational Supply Association
Primary school
Teaches matching of photographs to
    names in three languages

A46.   ANIMAL LOTTO; UK
–; Waddingtons Ltd
2 years old and over
2-4
Practice in colour and picture matching,
    and animal recognition

A47.   ANIMAL PUZZLE DOMINOES;
        UK
–; Abbatt Toys Ltd
3-7 years
Practice in shape and animal recognition

A48.   ANIMAL RACE; USA
P. McKee, M. L. Harrison; Houghton Mifflin;
    Abt Associates, Inc.
Children; pre-reading
2-4
Practice in matching pictures to initial
    consonants

A49.   ANIMAL TOSS; 1967; USA
H. A. Springle; Science Research Associates
Pre-school to 1st grade
1-4; 20-30 minutes

A50.   ANIMALS I; 1969; USA
H. A. Springle; Science Research Associates
Pre-school to 1st grade
1-4; 20-30 minutes

A51.   ANIMALS II; 1969; USA
H. A. Springle; Science Research Associates
Pre-school to 1st grade
1-4; 20-30 minutes

A52.   ANNE WILSON ACTION MAZE;
        1966; USA
A. A. Zoll; Addison-Wesley
Management
1
Insight into role of interpersonal relations in
    management

A53.   ANOTON; 1971; UK
–; Southampton Adult Education and Youth
    Service
Youth leaders
Leadership practice

A54.   ANZIO (Games 1, 2, 3, 4, 5 and 6);
        1969; USA
–; Avalon-Hill Inc
General
2+; 1-10 hours
Insight into Allied invasion of Italy in 1940

A55.   ANZIO BEACH HEAD; 1970;
        USA
–; Simulation Publications Inc.
General

A56. AO PRACTICE; 1962; USA
L. E. Allen; Wff'n Proof Inc
Grade 1 up to college
2+
Teaches use of disjunction-out rule of logic

A57. APEX; 1969; USA
R. Duke; University of Southern California
High school, college
Insight into air pollution

A58. APPLYING STRATEGIES AND
KNOWLEDGE I; 1969; USA
H. A. Springle; Science Research Associates
Pre-school to 1st grade
1-4; 20-30 minutes

A59. APPLYING STRATEGIES AND
KNOWLEDGE II; 1969; USA
H. A. Springle; Science Research Associates
Pre-school to 1st grade
1-4, 20-30 minutes

A60. APPOINTMENTS BOARD:
1971; UK
J. K. Jones; J. K. Jones
Secondary school
Insight into interviewing and appointment
problems

A61. APPRAISAL BY OBJECTIVES;
1970; USA
−; Didactic Systems Inc
College, management
3-5; 2-3½ hours
Practice in evaluating subordinates in
appraisal interviews and coaching to
improved performance

A62. THE ARIZONA BUSINESS
GAME; 1970; USA
J. A. Harlan, T. R. Navin, J. P. Logan;
Arizona University College of Business
and Public Administration
Graduate school, management
12-20; 30-48 hours
Practice in decision-making skills

A63. ARMADA; USA
−; KDI Instructional Systems Inc;
Educational Development Centre
Grade 8
Practice in planning strategy constrained
by political and physical factors

A64. ARMAGEDDON; USA
−; Simulations Publications Inc
General
Biblical warfare

A65. ARNHEM 1944; USA
−; Simulations Publications Inc
General

A66. ARTIFAX; UK
P.E. Consulting Group; P.E. Consulting
Group
College, management
Financial management

A67. AN ASPECT OF CHOICE;
1968; UK
D. Barber et al.; Methuen & Co Ltd
Secondary school
Insight into selecting a youth leader by
interview

A68. ASCOT (ANALOGUE
SIMULATION OF
COMPETITIVE
OPERATIONAL TACTICS);
1959; Canada
Imperial Oil Ltd; University of Toronto,
Department of Industrial Engineering
Management
2-15
Experience of decision-making in service
station management

A69. ASSISTANT SUPERINTENDANT
FOR BUSINESS MANAGE-
MENT; 1966; USA
D. Anderson, W. Hack; University Council
for Educational Administration
Educational administrators
Practice in problem-solving

A70.   ASSISTANT SUPERINTENDANT
       FOR INSTRUCTIONAL
       SERVICES; 1966; USA
B. Harris, R. Erskine; University Council
   for Educational Administration
Educational administrators
Practice in problem-solving

A71.   ASTRO-GUIDANCE
       PROGRAM EXERCISE; USA
A. A. Zoll; Addison-Wesley
Middle management
4 teams
Problem-solving exercise

A72;   ATLANTICA; USA
F. C. Davis; Albion
General
A 'Diplomacy' variant, set in 1872 bringing
   in the Confederate States, the Union and
   Canada

A73.   L'ATTAQUE; UK
—; H. P. Gibson Co Ltd
9 years old and over
2
Tactics

A74.   AUGUSTINE CITY; 1969; USA
R. Shukraft, J. Washburn, W. Relf;
   Simile II
High school
24-81; 3-48 hours
Initiation of programmes in church conflict
   context; testing of forms of lay-clergy
   ministries; increasing of sensitivity and
   motivation for church/community
   actions

A75.   AUSTERLITZ; USA
—; Simulations Publications Inc
General

A76.   AUTOCOUN; 1966; USA
J. W. Loughanz et al.; Personnel and
   Guidance Journal
High school, college, adults
Practice in 'automated counselling'

A77.   AUTOMOBILE DEALER
       MANAGEMENT DECISION-
       MAKING SIMULATION;
       1961; USA
J. Ruth, B. E. Spelder; Wayne State University
College, graduate school, management
4-48; 30 months
Principles of operating a dealership;
   experience in working with others to
   analyse and evaluate information and
   make decisions in a dynamic competetive
   situation

A78.   AVIATION; UK
—; H. Gibson
General
2; 2 hours
Aerial tactics game of attack and defence

A79.   AWKWORDS; UK
—; Waddingtons Ltd
10 years old and over
2-4
Spelling and vocabulary

A80.   THE AZIM CRISIS; 1971; UK
J. K. Jones; J. K. Jones
Secondary school
Insight into positions of statesmen and
   journalists in an international crisis

B1.   BALANCE; USA
—; Interact
7th grade, college
Up to 35
Insight into factors involved in balance in
   an eco-system

B2.   BANK LIQUIDITY MANAGE-
      MENT SIMULATION; 1969; USA
D. D. McNair, A. P. West; Simulated
   Environments Inc
Management
2-100; 1-2 days
Practice in the skills of managing sources
   and uses of bank funds

B3. BANK MANAGEMENT
SIMULATION (1401-FB-02X);
1963; USA
McKinsey Co Inc; IBM
Management, bank employees
3-45
Appreciation of relationships affecting
decisions or loans, deposits and investments, the need for planning to achieve
the most profitable use of assets; the
working relationship between the balance
sheet and the income statement, effects
of rising economy or recession

B4. BANKING; 1968; USA
E. Rausch, J. Cranmer; Didactic Systems
Inc and S.R.A. Inc
High school, college, graduate school and
management
3+; 2-4 hours
Insight into the need for profit-making by
banks and conscious influencing of
civic development

B5. BANKLOAN; USA
R. Rosen, C. Abt, R. Scott; Abt Associates
Inc
Bank management trainees
Insight into principles of loan-making

B6. BARBAROSSA; UK
−; Simulations Publications Inc
General
1941-45 war in Russia

B7. BASEBALL; USA
−; Sports Illustrated Games
General
2

B8. BASEBALL STRATEGY; USA
−; Avalon-Hill Inc
General
2
Insight into processes of team selection

B9. BASES BASIS; 1969; USA
R. Allen; Wff'n Proof
Primary school and upwards
Insight into number base

B10. BASIC BASIS; 1969; USA
R. Allen; Wff'n Proof
Primary school and upwards
Ordering in mathematics

B11. BASTOGNE; 1970; UK
−; Strategy and Tactics
General

B12. BATTLE OF BRITAIN GAME;
UK
−; Spiring Enterprises
General
2
Aerial combat

B13. BATTLE OF BRITAIN; USA
L. Zocchi; Simulations Publications Inc
General

B14. BATTLE OF BULGE
(GAMES 1 AND 2); 1965; USA
A. C. McAuliffe; Avalon-Hill Inc
General
2; 3-6 hours
Illustrates battle of Bastogne

B15. THE BATTLE OF BUNKER
HILL; 1971; UK
C. Grant; A. & C. Black
Military
4
Insight into the battle of Bunker Hill

B16. THE BATTLE OF MOLLWITZ;
1971; UK
C. Grant; A. & C. Black
Military
2
Insight into Battle of Mollwitz

B17. BATTLE OF MOSCOW; UK
−; Strategy and Tactics
General
German battle of Moscow 1941

B18. BATTLE OF STALINGRAD;
USA
−; Simulations Publications Inc
General
Strategy and tactics in World War II

B19.   BATTLE PLAN; 1972; USA
R. Loomis; R. Loomis
General
Computer moderated strategy game

B20.   BATTLE ROYAL; UK
—; Pepys Games
General
War game

B21.   BATTLESHIPS; UK
—; Waddingtons Ltd
8 years old and over
2
Tactical thinking

B22.   BAZAAR; USA
S. Sackson; 3M Co Ltd
Trading game

B23.   BEAT DETROIT; USA
—; Dynamic Design Industries
General
Running a car dealership

B24.   BEAT THE ELF; UK
—; Waddingtons Ltd
Primary school and upwards
1
Teaches topology

B25.   BELL ENGINEERING GROUP;
        1966; UK
J. Gill; Nelson Ltd
Management
Communication problems in a situation of
    rapid change

B26.   BICODE; UK
—; Griffin and George Ltd
Secondary school, college
Computer operations skills

B27.   BIG BOSS; 1972; Germany
—; I.W.A.
General
Business game

B28.   BIG LEAGUE; UK
—; Chad Valley Ltd
General
League football

B29.   BIG LEAGUE BASEBALL; USA
—; 3M Co Ltd
General

B30.   BIG LETTER FILL-UP; 1970;
        UK
—; Community Service Volunteers
Immigrants
Vocabulary practice

B31.   BINARY DOMINOES; 1972; UK
—; E. J. Arnold Ltd
Primary school
Practice in calculation

B32.   BINGOBANG
A. W. Heilman, R. Helmkamp, A. E. Thomas,
    C. J. Carsello; Lyons & Carnahan
Grades 1-3
2+
Practice in recognition of consonant sounds
    and symbols at the end of words; insight
    into concept of rhyming

B33.   BISMARK; 1967; UK
—; Strategy & Tactics
General

B34.   BISMARK; 1962; USA
—; Avalon-Hill Inc
General
20; 1-2 hours
Practice in strategic thinking and prediction

B35.   BISVIX; UK
V. J. Mills, Hatfield Polytechnic
Illustrates the attempted breakout situation
    of the Bismark during 1941

B36.   BISWICK AND DISTRICT
        TIMES; 1971; UK
G. Farrar; Community Service Volunteers
Secondary school
24
Insight into the problems of siting a
    hostel for the mentally sub-normal and
    techniques of problem solving

B37.   BLACKS AND WHITES; 1971; USA
R. Sommer, J. Tart; Psychology Today Games
High school, college, adults
Insight into racial aspects of house purchase as an illustration of the negro struggle for economic opportunity and political equality

B38.   BLACKBOARD GAME; 1945; UK
F. J. Schonell; Oliver & Boyd Ltd
Primary school
Reading practice

B39.   BLAST-OFF; UK
−; Waddingtons Ltd
9 years old and upwards
2-4
Space travel

B40.   BLENDS RACE; USA
A. W. Heilman, R. Helmkamp, A. E. Thomas, C. J. Carsello; Lyons & Carnahan
Grades 1-3
2+
Teaches fusion of letter sounds in two-letter blends

B41.   THE BLINDFORD GAME; 1969; USA
−; Training & Development Center
Management
2-16; 1-2 hours
Develops insight and sensitivity

B42.   BLITZKRIEG; USA
−; Avalon-Hill Inc
General
Practice in strategic and tactical thinking in context of Germany's breakout in World War II

B43.   BLOTS AND STRAIGHTS; 1972; UK
V. R. Parton; Albion
General
2+
Game based on the principle of alignment

B44.   BLOW FOOTBALL; Holland
−; Jumbo
General

B45.   BLOXBOX; 1972; Denmark
Piet Hein; Piet Hein
General
1
Seven cubes inside a transparent cube, aim is to achieve various colour and shape combinations

B46.   BLUE AND GREY; USA
−; Simulations Publications Inc
General
American civil war

B47.   BLUE LINE HOCKEY; USA
−; 3M Co Inc
General
2
Ice-hockey

B48.   BLUE WODJET COMPANY; USA
−; Interact
High school, college
25-30; 4-6 hours
Problems of pollution for industry

B49.   BMG; USA
−; Western Behavioural Sciences Institute
College, management
Insight into the function of competition in a large corporation

B50.   BODY TALK; USA
−; Psychology Today Games
College, adults
Up to 10; 10 minutes +
Insight into the use of bodies as an effective means of communication

B51.   BORODINO; USA
−; Simulations Publications Inc
General

B52.   BOSNIAN CRISIS; UK
W. van der Eyken; Brunel University
College
Insight into the 1909 crisis which led to the 1914-18 war

B53.   BREAKOUT; USA
—; Third Millennia Inc
General
World War II

B54.   BREAKOUT AND PURSUIT;
       USA
—; Simulations Publications
General
Insight into the Normandy campaign after
   the landings

B55.   BREAKTHRU; USA
—; 3M Company
General
2; 30 minutes
Board game on naval tactics

B56.   BRETWALDA GAME; UK
S.G.S. Associates; Longman Ltd
Primary, secondary school
History of Saxons and Vikings

B57.   BRINKMANSHIP: HOLOCAUST
       OR COMPROMISE; 1969; USA
D. Dal Porto; D. Dal Porto
High school
35; 2-4 days
Cold war crises (1945-1967); insight into
   difficulties of crisis between superpowers;
   role of national interest and security in
   foreign policy; negotiation under threat
   of nuclear war

B58.   THE BRIDGE GAME (PART
       OF A SUPERVISORY SKILLS
       SERIES); 1969; USA
—; Training Development Center
Management
3-15; 1-2 hours
Insight into planning and estimating,
   styles of supervision, delegation, manpower assignment, responsibility and
   morale

B59.   BRITISH STRATEGIC WAR
       GAME; 1905; UK
J. M. Grierson; War Office
Military
Insights into a possible German invasion
   of Belgium and its implications

B60.   BRITISH TRAINING AIRLINE;
       1963; UK
W. S. Barry, R. J. Crawford; British
   European Airlines
Management
3 teams
Practice in management of an airline

B61.   BROKEN LETTERS; USA
P. McKee, M. L. Harrison; Houghton
   Mifflin and Abt Associates Inc
Primary school (pre-reading)
1
Practice in discrimination of consonant
   letter sounds and forms

B62.   BUCCANEER; UK
—; Waddingtons Ltd
8 years old and over
2-4
Board game on piracy

B63.   BUDGETARY POLITICS; USA
L. Stitelman, W. D. Coplin; Science
   Research Associates
High school, college

B64.   BUDGETARY PROCESS;
       USA
L. Stitelman, W. D. Coplin; Science
   Research Associate
College, management

B65.   BUGS AND LOOPS; USA
—; Creative Specialities Inc
General
A programming game

B66.   BUILD; USA
A. J. Pennington; Drexel Institute of
   Technology
College, town planners
Insight into community development and
   the political, economic and social
   decisions involved

B67.   BUILDING A NEW TOWN;
       1971; UK
R. Walford; British Broadcasting Corporation
10-11 years old
1+
Insight into new town planning

B68. BULGE; USA
—; Simulations Publications Inc
General
Battle of the Bulge

B69. BUNDLING; USA
—; Plan B Corporation
Adults
Psychological game

B70. BUREAUCRACY; USA
—; Instructional Systems Inc
High school
Insight into the workings of bureaucracy

B71. THE BUS SERVICE GAME;
1969; UK
R. Walford; Longman
Secondary school
Insight into connected and dis-connected
network systems, urban hierarchies and
their effects on trade and traffic

B72. BUSHMAN EXPLORING AND
GATHERING; USA
R. E. Glazier; Abt Associates Inc
5th grade
Teaches concepts of cultural adaptation
to a harsh environment

B73. BUSINESS EXERCISE No. 1;
UK
—; International Computers Ltd
Management
Practice in decision-making in pricing,
marketing, research and development,
production, plant investment and
transport

B74. BUSINESS GAME; UK
—; Waddingtons Ltd
General
2-6; 1 hour +
Business decision making in mining and
transport

B75. THE BUSINESS GAME;
1966; UK
Imperial College; Department of
Management Science, Imperial
College
Students, management
Up to 6 groups; 20 ½ hour sessions
A management game covering production
management, marketing and financial
management in a company manufacturing
washing machines. Fortran based.

B76. THE BUSINESS GAME; UK
—; International Computers Ltd
College, management
25 minutes

B77. BUSINESS GAME FOR
USING ACCOUNTING
INFORMATION: AN
INTRODUCTION; 1965; USA
P. Fertig, D. F. Istvan, H. J. Mottice;
Harcourt, Brace, Jovanovitch
College, graduate school
2-8
Demonstrates the importance of financial
statements in decision-making

B78. BUSINESS GAMES–PLAY
ONE; 1958; USA
G. R. Andlinger; Harvard Business Review
College, management

B79. BUSINESS MANAGEMENT
DECISION SIMULATION
OF ORANGE COAST
COLLEGE; USA
—; Orange Coast College
College, management

B80. THE BUSINESS POLICY
GAMES; 1970; USA
—; Appleton-Century-Crofts
College, graduate school, management
1-20 hours
Insight into policy, strategy and the
integration of functions

B81.   BUSINESS SIMULATION; USA
R. H. Davies, J. A. Stephens; Westinghouse
Information Systems Laboratory
Management
15-40; 6-10 hours
Insight into variables and their interactions
  involved in key business decisions

B82.   BUX; 1966; USA
Western Behavioural Science Institute;
  Simile II
Junior high, high school, 9th grade, college
15-35; 3-8 hours
Teaches the concepts of inventory, sales,
  deficit financing, effect of advertising,
  research and development

B83.   BUY AND SELL GAME;
      1967; USA
H. A. Springle; Science Research Associates
Pre-school 1st grade
1-4; 20-30 minutes
Teaches concepts of addition and subtraction

B84.   BUYING GAME; 1968;
      South Africa
J. G. F. Wollaston; Greatermans Stores Ltd
College, management, buyers
5-15; 6-9 hours
Practice of inventory theory and its
  application

B85.   BYZANTIUM; USA
–; Rolling Front Game Co
General
Pattern making and breaking

B86.   BAFFLE; USA
P. Lamb; Lyons & Carnahan
Grades 1-6
2-7
Practice in construction of round patterns
  and words

C1.   CABBIE; 1973; UK
–; Intellect (UK) Ltd
General
Players compete to get their cab around
  London quicker than their opponents
  and to take the most money in fares
  and tips

C2.   CABINETS IN CRISIS; USA
–; WGBH-TV
High school
Insight into political processes

C3.   CABINETS IN CRISIS; 1968;
      USA
–; Foreign Policy Association
High school
15+; 2½-3 hours
Insights into the 1950 rejection by
  Yugoslavia of Soviet domination and
  into international decision-making

C4.   CADISIM (COMPUTER
      ASSISTED DISPOSAL
      SIMULATION); 1968; USA
L. C. Weir, D. A. Ameen; U.S. Army
  Logistics Management Center
Military
3-60; 3-5 hours
Illustrates management problems of large
  material depot and national inventory
  control point

C5.   CAGEY; USA
P. Lamb; Lyons and Carnahan
Grades 1-6
3-7
Practises vocabulary by 3, 4 and 5 letter
  word patterns

C6.   CAISIM (COMPUTER
      ASSISTED INDUSTRY
      SIMULATION); 1962; USA
J. Arnett, W. Ketner, M. J. McGrath;
  U.S. Army Logistics Management
  Center
Military
3-60; 2-4 hours
Illustrates problems of industrial
  production management

C7. CALOGSIM (COMPUTER
ASSISTED LOGISTICS
SIMULATION); 1959; USA
U.S. Army Logistics Management Center;
U.S. Army Logistics Management Center
Military
6-144; 24 hours
Decision making and supply management;
analysis of complex information; insight
into inter-relation of management areas.

C8. CALTECH POLITICAL
MILITARY EXERCISE; USA
−; California Institute of Techology
College, military

C9. CAMPAIGN; 1970; USA
−; Instructional Simulations Inc
Junior high school, college, adults
16-32; 4-8 hours
Insight into a state leglislature election in a
two-party system

C10 CAMPAIGN; UK
−; Waddingtons Ltd
General
2-4
Insight into Napoleonic wars and
alternative strategies

C11. CAMPUS CRISIS GAME; USA
WGBH Educational Foundation, WHYY
Educational Television; WGBH-TV
Adults
10+; 90 minutes
Insight into causes of campus riots and
strikes

C12. CAMSIM (COMPUTER
ASSISTED MAINTENANCE
SIMULATION); 1962; USA
M. J. McGrath; US Army Logistics
Management Center
Military
3-60; 2-4 hours
Illustrates problems of managing a maintenance
activity

C13. CANDIDATES, ISSUES
AND STRATEGIES; 1964;
USA
I. Pool, De Sola, R. P. Abelson, S. Poplin;
M.I.T. Press
High School
1960 and 1964 presidential elections;
insight into voter behaviour

C14. C & O/B & O; 1969; USA
Avalon-Hill; Didactic Systems Inc
General
Railway scheduling game; insight into
problems and processes of running
railways

C15. CAPERTSIM (COMPUTER
ASSISTED PROGRAM
EVALUATION AND REVIEW
TECHNIQUE SIMULATION);
1962; USA
US Army Logistics Management Center;
US Army Logistics Management Center
Military
4-84; 4-6 hours
Experience of evaluating cost-time
relationships and trade-offs, evaluating
alternative strategies decision-making
in uncertainty, use of PERT and use of
a computer as a simulator

C16. CAPSTONE; USA
Insite Project; Indiana University
Education students, teachers
Insight into the problems of the beginner
teacher

C17. CAPTURE; USA
P. Lamb; Lyons and Carnahan
Grades 1-6
3-4
Discrimination of syllables in words

C18. CARAPACE; USA
−; Plan B Corporation
General
Tactical board game

C19. CAR-CAPERS; UK
—; Spears Games
6 years old and over
2-6
Visual discrimination

C20. CARD CRICKET; 1972; UK
G. Levin, A. Witkin; Games & Puzzles
General
Cricket knowledge

C21. CAREERS; USA
—; Parker Bros
General
2-6
Board game on progress through certain
careers

C22. CARESIM (COMPUTER
ASSISTED REPAIR
SIMULATION); 1963; USA
US Army Logistics Management Center;
US Army Logistics Management Center
Military
3-60; 2-3 hours
Insight into problems of establishing sound
maintenance policies on replacement of
parts in equipment and weapon systems

C23. CARGO GAME; USA
—; Edcom Systems Inc; Abt Associates Inc
4th grade
Insight into decisions made by Zinacanteco
Indians of Mexico

C24. CARIBOU HUNTING: BOW
AND ARROW HUNTING
GAME; 1969; USA
H. Kinley; Curriculum Development
Associates; MACOS
Upper elementary grades
3; 10-20 minutes
Hunting strategy of Netsilik Eskimoes

C25. CARIBOU HUNTING:
CROSSING PLACE HUNTING
GAME; 1969; USA
H. Kinley; Curriculum Development
Associates; MACOS
Upper elementary grades
3; 10-20 minutes
Hunting strategy of Netsilik Eskimoes

C26. CARMSIM (COMPUTER
ASSISTED RELIABILITY
AND MAINTAINABILITY
SIMULATION); 1966; USA
W. D. Ketner; US Army Logistics
Management Center
Military
3-60; 1½-2 hours
Illustration of the application of computer
to a cost/reliability problem

C27. CARNEGIE MANAGEMENT
GAME; 1957, 1964; USA
P. Winters, A. Kuehn, W. Dill, K. Cohen;
Graduate School of Industrial
Administration, Carnegie-Mellon
University
Graduate school
15+; 12 months
A total management game modelled on
US detergent industry

C28. CASE A: FACTORY LEVEL;
1970; UK
T. T. Patterson; University of Strathclyde
Personnel managers
Insight into problems of industrial relations

C29. CASE B: NATIONAL
LEVEL; 1970; UK
T. T. Patterson; University of Strathclyde
Personnel managers
Insight into problems of industrial relations

C30. CAVEAT EMPTOR; USA
—; Plan B Corporation
General
Financial strategy

C31. CENTURION; UK
—; Strategy & Tactics
General
Roman warfare

C32. CHAMPIONSHIP GOLF; USA
—; Championship Games
General

C33. CHANCELLORSVILLE; 1967;
UK
—; Strategy & Tactics
General
US Civil War

C34. CHAQUICOCHA; 1972; UK
J. P. Cole, P. M. Mather, P. T. Whysall;
Department of Geography, Nottingham
University
Secondary school upwards
Geographical game: man against the
environment in an Andean agricultural
settlement. Can be played on a computer,
thus also teaches programming skills

C35. CHARTBUSTER; 1973; UK
—; ASL Pastimes
General
Board game on pop music business

C36. CHASE; UK
—; Rupert Hart-Davis Ltd
Primary school
Teaches word recognition

C37. CHATTER; 1973; UK
—; Intellect (UK) Ltd
General
Players compete in the collection and then
translation of foreign words and phrases
(French and Spanish)

C38. CHECK AND CHANGE; 1972;
UK
—; E. J. Arnold Ltd
Primary, secondary school
Use of decimal currency skills

C39. CHECKS AND BALANCES;
1969; USA
D. Dal Porto, R. Lundstedt; Mount Pleasant
High School
High school
25-140; 5-6 hours
Insight into New Deal history and the
mechanics of USA form of government

C40. CHECKLINES; UK
—; Triang Ltd
Primary school
2
Spatial relationships

C41. CHEMSYN; 1972; UK
G. Eglinton, J. R. Maxwell;
Heyden & Son
Secondary school and upwards
Chemical card game, cards giving details of
50 basic compounds from a range of
classes belonging to aliphatic and
aromatic organic chemistry

C42. CHESSWORD; UK
—; Waddingtons Ltd.
8 years old and over
2
Tactical thinking, spelling and vocabulary

C43. CHICAGO, CHICAGO! USA
—; Simulations Publications Inc
General
1968 street conflict in Chicago

C44. CHINESE CHECKERS;
Denmark
—; Piet Hein
General

C45. CHINESE TAC-TICKLE;
1965; USA
H. D. Ruderman; Wff'n Proof Inc
Primary school and upwards
2
Strategic thinking

C46. CHUG A LUG; USA
—; Dynamic Design Industries
-General
Game about drinking

C47. C1 PRACTICE; 1962; USA
L. E. Allen; Wff'n Proof Inc
Grade 1, college
2+
Use of implication in rule of logic

C48. THE CIRCUIT GAME; 1971
UK
J. Bloomer; Department of Education,
Glasgow University
Secondary school
5+ minutes
Circuitry, basic concepts and principles of
electricity

C49. CIRCUS DOMINO; Holland
D. Bruna; Jumbo Games
Children

C50. CIRCUS JUMBOLI; Holland
−; Jumbo Games
Primary school
Insight into life at circus

C51. THE CITIES GAME; USA
−; Psychology Today Games
General
The processes, relationships, groups, etc
involved in city government, in terms of
urban collapse and renewal

C52. CITY I; USA
−; Envirometrics
High school, college, graduate school,
management
Insight into working of an urban system
in terms of money, land and political
power

C53. CITY II; USA
−; Environmetrics
High school, college, graduate school,
management
20-70; 8+ hours
Insight into working of an urban system in
terms of finance, law, political power,
transportation, immigration and social
factors

C54. CITY III; USA
−; Environmetrics
High school, college, graduate school,
management
Insight into working of a metropolitan area
system in terms of finance, land, politics,
immigration, transportation and other
social factors

C55. CITY MODEL; USA
−; Applied Simulations International Inc
College, management
30-150; 3-20 hours
To increase level of awareness to achieve a
holistic outlook and an interdisciplinary
approach to problem solving

C56. CIVIL DISORDER GAMING;
USA
R. J. Butchers, R. M. Longmire; Research
Analysis Corp
Police
3-10; 3 hours
Increase police readiness for dealing with
civil disorder

C57. CLICK; UK
−; Waddingtons Ltd
10 years old and over
2-4
Teaches number and four rules

C58. THE CLUB GAME; UK
A. C. Daubney; Hastings Youth Offices
Youth leaders
Leadership decision-making in youth clubs

C59. CLUEDO; UK
−; Waddingtons Ltd.
General
3-6
Detection, deduction

C60. CLUG (COMMUNITY LAND
USE GAME); 1966; USA
A. G. Feldt; Systems Gaming Associates
High school, college
3-25; 3-20 hours
Teaches classical principles of urban and
regional economics

C61. CLUG (CORNELL LAND
USE GAME); 1965; USA
−; Cornell University
University
Classical principles of urban and regional
economics

C62.   CLOTHING I; 1969; USA
H. A. Springle; Science Research Associates
Pre-school to grade 1
1-4; 20-30 mins.

C63.   CLOTHING II; 1969; USA
H. A. Springle; Science Research Associates
Pre-school to grade 1
1-4; 20-30 mins.

C64.   COBRA (COMPUTER BLAST
       AND RADIATION
       ASSESSMENT); USA
—; Army Strategy & Tactics Analysis
   Group
Military
Insight into aftermath of strategic nuclear
   war

C65.   COKERHEATON STORY
       PACK; UK
—; Evans Bros
9-14 year olds
Improve English through story set in an
   industrial town

C66.   COLCHESTER EXERCISE
       1968; UK
M. Nicholson, L. Suransky; BBC
General
Insight into Arab-Israeli conflict

C67.   COLD WAR; 1972; UK
R. Blackshaw; Albion
Adults
Insight into the cold war of 1950s

C68.   COLD WAR; USA
—; Simulations Publications Inc
Adults

C69.   COLD WAR GAME; 1954; USA
H. Goldhamer; Rand Corporation
Politico-military
Examination of alternative strategies

C70.   COLLECTIVE BARGAINING;
       USA
E. Rausch; Didactic Systems Inc
Junior and middle management
6+
Insight into political and economic forces
   and the need for careful semantics

C71.   COLLECTIVE BARGAINING;
       1968; USA
E. Rausch; Science Research Associates
High school, college
6+; 2-5 days
Insight into collective bargaining and skills
   of negotiating and administrating union
   contracts

C72.   COLLEGE FOOTBALL; USA
—; Sports Illustrated Games
General
2

C73.   COLLEGE GAME; 1968; USA
R. R. Short; Whitworth College
High school, college.
3+; 2-2½ hours
Freshman orientation, communication
   between students and faculty

C74.   COLLEGE AND UNIVERSITY
       PLANNING GAME; 1963; USA
J. Forbes; New Mexico State University
University
Value classification underlying academic
   growth

C75.   COLONY; 1970; USA
A. K. Gordon; Science Research Associates;
   Abt Associates Inc
Secondary school, high school grade 8,
   junior college
Pre-revolution relationship between the
   USA and the UK

C76.   COLOR RHYMING LOTTO;
       1969; USA
A. R. Coller, L. G. Gotkin; New Century
Primary school
3-5; 15-20 minutes
Practice in matching names to colours,
   verbal descriptions and recognition of
   verbal descriptions

C77. COLOR; USA
M. S. Gordon; Abt Associates Inc
Elementary school
Practice in colour mixing from primary
   colours

C78. COLOUR DOMINOES; UK
—; Spears' Games
Primary school
Teaches number concepts

C79. COLOUR SNAP; UK
—; Orchard Toys Ltd
Primary school
Colour recognition

C80. COMBAT; UK
B. Horrocks; Merit Ltd
10 years old and over
2-4
Practice of strategic and tactical thinking

C81. COMEXOPOLIS; 1966; USA
R. D. Duke, J. D. Reske; University of
   Southern California
University
Air pollution

C82. COMMERCIAL CREDIT
   SIMULATOR; 1969; USA
D. D. McNair, A. P. West; Simulated
   Environments Inc
Management
2-40; 2-6 hours
Insight into and practice of commercial
   lending and portfolio managements

C83. COMMUNICATION; USA
—; Education Research
College, management
Communications in business

C84. THE COMMUNITY; 1968; USA
E. Rausch, H. J. Cranmer; Science Research
   Association; Didactic Systems Inc
High school, college, management
3+; 2-5 hours
Illustrates basic problems of selecting and
   financing public services

C85. COMMUNITY COLLEGE
   PRESIDENCY; 1966; USA
L. Johnson; University Council for
   Educational Administrators
Educational administrators
Practice in conflict resolution

C86. COMMUNITY DISASTER;
   1970; USA
M. Inbar; Diadactic Systems Inc; Western
   Publishing Co Inc; Academic Games
   Associates Inc
Junior high, high school, college, adults
6-16; 2-6 hours
Insight into the conflict between anxiety
   about family and responsibility to the
   community

C87. COMMUTATIVE TAC-TICKLE;
   1967; USA
H. D. Ruderman; Wff'n Proof Inc
Primary school and upwards
Practice of strategic thinking and insight
   into commutative process

C88. COMPLETE WORLD WAR II;
   USA
—; Guidon Games
General

C89. COMPUMAN; 1969; USA
J. A. Craven; Graduate School of Business
   Administration, University of Washington
College, graduate school, management
2-9
Practice of business theory

C90. COMPUTAMATIC BASEBALL;
   USA
—; Electronic Data Controls
General
2

C91. COMPUTAMATIC BASKETBALL;
   USA
—; Electronic Data Controls
General
2

C92. COMPUTAMATIC FOOTBALL; USA
—; Electronic Data Controls
General
2

C93. COMPUTAMATIC GOLF; USA
—; Electronic Data controls
General
4

C94. COMPUTAMATIC HOCKEY; USA
—; Electronic Data Controls
General
2

C95. COMPUTAMATIC SAILING; USA
—; Electronic Data Controls
General
2

C96. COMPUT-A-TUTOR; USA
—; Didactic Systems Inc
General
Insight into computer programming

C97. COMPUTER CITY; 1970; USA
—; Applied Simulations International
College, management
9-45; 3-15 hours
Increase awareness to achieve holistic
outlook and interdisciplinary approach
to problem-solving

C98. COMPUTER SALES GAME; USA
—; R.C.A.
Management

C99. COMPUTER TIC-TAC-TOE; UK
—; The Chad Valley Co
General

C100. CONDUCTING PLANNING
EXERCISES; USA
P. A. Twelker; Instructional Developments
Corporation
High school
16-36

C101. CONDUR; UK
—; Management Games Ltd
'A'-level and upwards
1-24
Insight into the marketing a consumer
durable product

C102. CONEX; 1967; UK
M. H. Banks, A. J. R. Groom,
A. N. Oppenheim; Edinburgh
University
Political
20-30
Insight into a Middle East crisis

C103. CONFETTI THE CLOWN; UK
—; Merit Ltd
Primary school
Colour perception

C104. CONFIGURATIONS:
DESARGUES 10_3;
1967; USA
H. L. Dorwart; Autotetic Instructional
Materials Publishers, Wff'n Proof,
International Learning Corp
General
1

C105. CONFIGURATIONS: THE
FANO 7_3; 1967; USA
H. L. Dorwart; Autotetic Instructional
Materials Publishers, Wff'n Proof,
International Learning Corp
General
1
Insight into the geometry of incidence

C106. CONFIGURATIONS: THE
MÖBIUS-KANTOR 8_3; 1967;
USA
H. L. Dorwart; Autotetic Instructional
Materials Publishers; Wff'n Proof;
International Learning Corp
General
1
Insight into the geometry of incidence

C107.   CONFIGURATIONS: THE
PAPPUS 9_3; 1967; USA
H. L. Dorwart; Autotetic Instructional
Materials Publishers; Wff'n Proof;
International Learning Corp
General
1

C108.   CONFLICT; USA
−; World Law Fund
Adults
24-36; 2-3 hours
International peacekeeping and policing

C109.   CONFRONTATION; 1971;
USA
G. Scott; Creative Communications and
Research
Students
4+
Protest and the Establishment

C110.   CONFRONTATION; 1967; USA
−; Gamescience
General
Psychological game

C111.   CONFRONTATION; 1967; USA
−; Simulations Publication
General
World strategy game

C112.   CONGRESSIONAL
COMMITTEES; USA
L. Stitelman, W. D. Coplin; Science Research
Associates
High school
Working of US Congress

C113.   CONGRESSMEN AT WORK;
USA
L. Stitelman, W. D. Coplin; Science
Research Associates
High school
Working of US Congress

C114.   CONNECT; 1972; UK
K. Garland; Galt Toys Ltd
5 years old to adult
Pattern recognition

C115.   CONSERVATION; 1972; UK
−; Royal Society for the Protection of
Birds
General
2-4; 1 hour
Insight into relationship between pollution,
protection and conservation; knowledge
of bird species and habitats

C116.   CONSONANT BINGO; 1970
UK
−; Community Service Volunteers
Immigrants
3+
Phonic practice

C117.   CONSONANT LOTTO;
1956; USA
E. W. Dolch; Garrard Publishing Co
First grade and upwards
2-8; 10-15 minutes
Sound and consonant matching

C118.   CONSTRUCTIVE
DISCIPLINE; 1969; USA
−; Didactic Systems Inc
College, management
3+; 2-3½ hours
Practice in handling situations which
influence discipline and morale

C119.   CONSUMER; 1969. USA
G. Zaltman; Western Publishing Co Inc;
Didactic Systems Inc; Academic Games
Associates Inc
Junior high school, adults
11-34; 1½-2½ hours
Problems and economics of instalment
buying

C120.   CONSUMER GAME; 1966; USA
G. Zaltman; Board of Co-operative
Educational Services
Junior high, high school
1; 1-2 hours
Complexities of hire purchase, calculation
of true interest rates, benefits of
budgeting

C121.   CON-TAC-TIX; Denmark
—; Piet Hein
General

C122.   CON-TAC-TIX; USA
—; Didactic Systems Inc
Management
5-8
Insight into task group processes

C123.   CONTINENTAL; 1947; UK
P. Adolph; Subbuteo Ltd
General

C124.   CONTINGENCY RESOURCE
        ALLOCATION; 1970; UK
R.H.R. Armstrong, M. Hobson; Institute of
        Local Government Studies, Birmingham
        University
Post-experience, post-graduate, local
        government
4+
An application of NEXUS (q.v.) to study of
        CRA

C125.   CONTRABAND; UK
—; Castell
General
3+
Smuggling game

C126.   CONTRACT NEGOTIATIONS;
        1970; USA
J. Zif, R. E. Otlewski; Macmillan Co
College, management
8-15; 3-6 hours
Insight into strategy of labour negotiations
        and collective bargaining and development
        of negotiating skills

C127.   CONTROL IN SCHOOL;
        1972; UK
School Council; Penguin Books
Education students, teachers secondary
        school
Insight into conflict in secondary schools

C128.   CONTROLLED PACE
        NEGOTIATION; UK
N. Rackham; Industrial and Commercial
        Training Negotiating game

C129.   CONVENTION!; 1960; USA
H. Babbidge; Games Research Inc
General
3-7; 2-4 hours
Workings of presidential convention

C130.   CO PRACTICE; 1962; USA
L. E. Allen; Wff'n Proof Inc
Grade 1, college
2+
Teaches use of implication-out rule of
        logic

C131.   THE CORK BALLS GAME;
        1969; USA
—; Training Development Center
Management
4-16; 1-2 hours
Teaches principles of work simplification,
        preparation and use of flow-charts and
        diagrams

C132.   CORN GAME; USA
—; Edcom Systems Inc; Abt Associates Inc
4th grade
Insight into the decisions made by the
        farming Zincanteco Indians of Mexico

C133.   CORNELL HOTEL
        ADMINISTRATION
        SIMULATION EXERCISE;
        1968; USA
R. M. Chase; School of Hotel Administration,
        Cornell University
College, management
12-32; 4-8 hours
Insight into functional decision area in
        hotel and managerial economics

C134.   CORNELL RESTAURANT
        ADMINISTRATION
        SIMULATION EXERCISE;
        1970; USA
R. M. Chase; School of Hotel Administration
        Cornell University
College, graduate school, architects,
        management
12-32; 6-8 hours
Integration of major functional areas into
        single strategy

C135.   CORPS GAME; UK
—; Defence Operational Analysis Establishment
Military
6-8 weeks
Insight into war strategy and tactics when
heavily outnumbered

C136.   CORRIDOR; 1967; USA
M. S. Gordon, J. C. Hodder; Abt Associates
Inc
High school
Insight into technical, economic and political
constraints on alternative plans for
transportation in North East Corridor

C137.   COTTAGE GAME; 1972; UK
—; Habitat Ltd
4 years old and over
Early Victorian 'morality game'

C138.   COUNT A WFF; 1962; USA
L. E. Allen; Wff'n Proof Inc
Grade 1 to college
2+
Practice in recognition and construction of
well formed formulae

C139.   COUNTMASTER MAJOR;
1971; UK
J. Hicks, T. Kremer; MacDonald Ltd
Primary school
Practice in relating length to figures,
addition, comparison and subtraction

C140.   COUP D'ETAT; UK
—; Miniature Warfare
General
Tactics and strategy of a coup

C141.   CRASH; UK
—; Matchbox Toys
General
Grand Prix motor-racing

C142.   CRAZY CATERPILLAR; UK
—; Waddingtons Ltd
3 years old and over
Any number
Colour and shape-matching and recognition

C143.   CREATIVE BASIS; 1969; USA
R. Allen; Wff'n Proof Inc
Primary school and upwards
Ordering and number bases

C144.   CREDIT UNION MANAGE-
MENT; 1968; USA
H. E. Thompson; IBM; CUES Managers
Society
Management
1-699
Training credit union executives in loan
procedures, cash management and rates
of loans

C145.   CRETE; 1969; USA
—; Simulations Publications Inc
General
Battle of Crete

C146.   CRICKET GAME; 1969; UK
—; Waddingtons Ltd
7 years old and over
2
Teaches rules of cricket, fielding positions,
etc.

C147.   CRISIS; 1966; USA
—; Western Behavioural Sciences Institutes;
Didactic Systems Inc; Simile II
Grade 6 to graduate school
18-36; 50 minutes +
Insight into conflict over mining a rare
element, interaction of nations and world
organisations

C148.   CRISIS: CONGO; 1968; USA
R. Clarke, S. Smith, W. McQueeney;
WGBH Educational Foundation
High school
15+; 2½-7½ hours
Concepts, attitudes and strategies of decision-
making, analysis of international events,
effect of national interest on international
alliances

C149.   THE CRISIS GAME; 1965;
USA
S. F. Giffin; Doubleday
Simulation of international conflict

C150.   A CRISIS GAME FOR
        ECONOMIC STUDIES; 1970;
        West Germany
H. Steffens; Pädagogische Beiträge
Secondary school
2
Insight into and practice of problem solving
   in economic development of countries at
   starvation point

C151.   CRISIS IN LAGIA; UK
J. Elliot; Schools Council/Nuffield
   Humanities Project
Secondary school
Insight into aspects of war and society

C152.   CRISISCOM; USA
–; Massachusetts Institute of Technology
Adult
Practice in information processing in an
   international context

C153.   CRISS CROSS; USA
P. Lamb; Lyons and Carnahan
Grades 1-6
2-7
Word building by phoneme-grapheme
   relationships

C154.   CROSS COUNTRY; USA
P. Lamb; Lyons and Carnahan
Grades 1-6
3-8
Recognition of phoneme-grapheme
   relationships by diacritical marks and
   phonetic spelling

C155.   CROSS-NUMBER PUZZLES–
        DECIMALS AND PERCENT;
        1968; USA
M. Murfin, J. Bazelon; Science Research
   Associates
10-15 years old
1-40
Arithmetical exercises based on crossword
   puzzle

C156.   CROSS-NUMBER PUZZLES–
        FRACTIONS; 1967; USA
M. Murfin, J. Bazelon; Science Research
   Associates
9-13 years old
1-40
Arithmetical exercises based on crossword
   puzzle

C157.   CROSS-NUMBER PUZZLES–
        STORY PROBLEMS; USA
M. Murfin, J. Bazelon; Science Research
   Associates
10-15 years old
1-40
Arithmetical exercises based on crossword
   puzzle

C158.   CROSS-NUMBER PUZZLES–
        WHOLE NUMBERS; 1966; USA
J. Murfin, J. Bazelon; Science Research
   Associates
7-12 years old
1-40
Arithmetical exercises based on crossword
   puzzle

C159.   CROSS OVER; USA
P. Lamb; Lyons and Carnahan
Grades 1-6
2-8
Vocabulary building

C160.   CRUMOX; UK
P.E. Consulting Group; P.E. Consulting
   Group
College, management
2+
Management experience in market policy
   testing and in control of cash and stocks

C161.   CRUX; 1972; Denmark
Piet Hein; Piet Hein
General
1
Solid cross with 6 projecting, separately
   rotating arms; aim is to bring three spots
   of different colours together at each
   intersection

C162.　CRYPTODIPLOMACY; UK
J. C. McCallum; Albion
General
7
A postal version of 'Diplomacy' using coded
　messages

C163.　CUBE FUSION; UK
—; Waddingtons Ltd
General
2
Practice of strategic thinking.

C164.　CULTURE CONTACT; USA
C. Isber, R. Glazier; Games Central
12-18 years old
20-30
Insight into conflict of cultural relativism

C165.　CUTTING GRASS; USA
—; Instructional Simulations Inc
Grades 7-12
Insight into peer pressure tactics and drug
　problems

C166.　CZAR POWER; USA
—; Instructional Simulations Inc
9-12 Grade
Insight into autocracy

D1.　DAISY GAME; UK
—; Spears Games
4 years old and over
2-4
Colour matching and discrimination

D2.　DANGEROUS PARALLEL;
　1969; USA
R. G. Mastrude; Scott, Foresman Co; Abt
　Associates; Foreign Policy Association
Junior high, high school, college
18-36; 5 hours
38th parallel in Korean War; insight into
　factors affecting foreign policy decision
　making

D3.　DARK AGES; USA
—; Simulations Publications Inc
General
Tactics of Vikings, Franks etc. in Byzantine
　period

D4,　DATACALL; 1970; USA
R. R. Johnson; Department of Psychology,
　Earlham College
College
1-70; 1-2 hours +
Decision processes in experimental design

D5.　DAYTON TYRE SIMULATION;
　1959; USA
P. S. Greenlaw; American Marketing
　Association
Management
12-24; 3 months
Practice of skills in dynamic, interactive
　decision making

D6.　D-DAY; 1965; USA
—; Avalon-Hill Inc
General
2+; 2-6 hours
World War II from Normandy landings;
　practice of strategic and predictive
　thinking

D7.　DECIMAL DOMINOES: 1970;
　UK
—; Waddingtons Ltd
General
Practice in use of decimal currency

D8.　DECIMAL MONEY DOMINOES;
　1972; UK
E. J. Arnold Ltd
Primary, secondary school
Practice in decimal currency and number
　rules

D9.　DECISION; USA
—; Education Research
College, management
Business decision-making

D10.　DECISION GAME; UK
—; National Association of Youth Clubs
Youth leaders
Decision-making leading to a successful
　youth club

D11.   DECISION MAKING; 1967; USA
E. Rausch; Didactic Systems Inc; Science
Research Associates
College, graduate school, management
3+; 1½-2½ hours
Development of human resources,
   management philosophy, matching
   resources to tasks, consistency in
   planning

D12.   DECISION MAKING; USA
L. Stitelman, W. D. Coplin; Science
Research Associates
Decision-making

D13.   DECISION MAKING IN
      ELEMENTARY SCHOOL; USA
R. Wynn; University Council for Educational
Administrators
Educational administrators

D14.   DECISION-MAKING
      SIMULATION; 1966; UK
N. Cuthbert; Nelson Ltd
Management
Personal and organisational relationships
   under stress

D15.   DECISION SIMULATION OF
      A MANUFACTURING FIRM,
      USA
B. H. Sord; College of Business Administration, University of Texas
Graduate school
4+; 1 hour
Insight into operation and management of
   a production company

D16.   DECISIONS I: SITING AN
      OIL TERMINAL; 1971; UK
C. R. A. Atthill; School Government
Publishing Co
Sixth forms, college
16; 19 hours
Use of planning data in decision-making in
   industry; economic, social, marketing
   and environmental factors

D17.   DECISIONS II: COASTAL
      TANKERS; 1972; UK
C. R. A. Atthill; School Government
Publishing Co
Sixth form, colleges
16; 19 hours
Insight into operational, economic and
   personnel decisions in industry

D18.   DECISIONS III: MARKETING;
      1973; UK
C. R. A. Atthill; School Government
Publishing Co
Sixth form, college
16; 19 hours
Marketing decision-making

D19.   DECLINE AND FALL; UK
—; Wargames Research Group
General
4
Barbarian invasions and fall of Rome
   375-450 AD

D20.   DE KALB POLITICAL
      SIMULATION; 1969; USA
W. Harader, C. N. Smith, M. H. Whithed;
   Political Science Department, Rensselaer
   Polytechnic Institute
High school, college
28-77; 6-8 hours
Dynamics of an electoral campaign

D21.   DELEGATION; USA
—; Education Research
College, management
Delegation in business setting

D22.   DELPHI I (A COMPUTER-
      BASED EXPLORATION OF
      ALTERNATIVE FUTURES);
      1968; USA
—; Computer Based Education Research
   Laboratory, University of Illinois
College
1-18; 1-1½ hours
Enlargement of perspectives on world
   problems

D23.   DEMOCRACY; USA
—; 4H Foundation
High school
6+; 30 minutes
Legislative and community meetings

D24.   DEMOCRACY; USA
J. S. Coleman; Western Publishing Co
High school
7-8; 30 minutes
Practice in group decision-making in
    government and political science

D25.   DEMOCRACY; 1969; USA
J. S. Coleman; Western Publishing Co
Junior high, high school
6-11; ½-4 hours
Teaches necessity of negotiation and exchange,
    natural origins of 'logrolling' and
    satisfying one's constituents

D26.   DERBYSHIRE; UK
—; Derbyshire Youth Service
Youth leaders
Club leaders 'in-tray exercise'

D27.   DESTINY; 1969; USA
P. De Kock, D. Yount; Interact
Grade 7, college
25-40; 15-30 hours
Practice in differentiating fact, inference and
    judgement set in Cuban crisis of 1898;
    insight into Presidential decisions

D28.   DESTRUCTION OF ARMY
        GROUP CENTRE; USA
—; Simulations Publications Inc
General, military
Warfare in Russia in 1944

D29.   DETECTIVE; UK
—; Inter-Action
General
Practice in logical analysis

D30.   DEVELOPMENT; 1970; USA
H. Kaplan; Science Research Associates;
    Abt Associates Inc
High school, junior college
Insight into major powers' courting of
    neutral countries' favour by aid in
    return for support in international crises

D31.   DEVELOPING COUNTRIES;
        1970; UK
R. H. R. Armstrong, M. Hobson; Institute
    of Local Government Studies, Birmingham
    University
Post-experience, post-graduate, local
    government
4+; 2 hours +
Application of NEXUS (q.v.) to study of
    problems of developing countries

D32.   THE DEVELOPMENT GAME;
        1969; UK
R. Walford; Longman
Secondary school
Insight into problems of development in
    underdeveloped areas and practice in
    search techniques

D33.   THE DEVELOPMENT GAME—
        BRITISH HONDURAS; 1970; UK
O. G. Thomas, M. F. C. Clarke; Oxfam
Secondary school, adults
24-60
Insight into development planning priorities
    and conflicts between givers and takers

D34.   DEVIL'S ADVOCATE GAME;
        UK
SGS Associates; Longman
Primary school and upwards
History of Saxons and Vikings

D35.   DIADOKHI; UK
G. Jeffrey; G. Jeffrey
Diplomacy variant

D36.   DIADOKHI II; UK
G. Jeffrey; G. Jeffrey
General
Diplomacy variant

D37.   DICK BRUNA CARD
        GAME 4; Holland
—; Jumbo Games
Primary school

D38.   DICK BRUNA LOTTO; UK
D. Bruna; Educational Supply Association
Primary school
Up to 6

D39.   DIG; 1969; USA
J. Lipetsky; Interact
Fifth grade, college
30-40; 15-30 hours
Insight into inter-relationships of cultural
    patterns

D40.   DIGRAPH HOPSCOTCH; USA
A. W. Heilman, R. Helmkamp, A. E. Thomas,
    C. J. Carsello; Lyons and Carnahan
Grades 1-3
2+
Recognition of digraph sounds and symbols

D41.   DIGRAPH WHIRL; USA
A. W. Heilman, R. Helmkamp, A. E. Thomas,
    C. J. Carsello; Lyons and Carnahan
Grades 1-3
2+
Recognition of initial consonant digraph
    sounds and symbols

D42.   DIPLOMACY; 1960; USA
A. Calhammer; Philmar Ltd; Games
    Research Inc
College, military
4-7; 3-6 hours
Insight into the processes of diplomacy among
    the great powers in 1900 to 1914

D43.   DIPLOMYOPIA; UK
C. Hemming; XL
A variant of 'Diplomacy'

D44.   DIRTY WATER; 1970; USA
J. Anderson, H. Trilling, R. Moody, R. Rosen;
    Urban Systems Inc
8 years old and over, adults
2-4; 1-2 hours
Decision making to deal with pollution

**D45.   DISASTER (COMMUNITY
    RESPONSE); USA**
−; John Hopkins University
High school
Problems and behaviour of a community
    under disaster conditions

D46.   DISCLOSE; 1971; UK
J. R. Cooper; J. R. Cooper, Dunchurch
    Industrial Staff College
University, management
1½ days +
A manufacturing situation, focusses on
    importance of pre-planning for production
    control and cost and quality control

D47.   DISCOLLATE; 1972; UK
J. R. Cooper; J. R. Cooper; Dunchurch
    Industrial Staff College
University, management
1 day
Production scheduling; worker motivation
    through job enlargement

D48.   DISORGANISE; 1972; UK
−; Dunchurch Industrial Staff College
University, management
1 day
Manpower planning, organisation planning

D49.   DISCOVERY; 1972; UK
−; Dunchurch Industrial Staff College
University, management
1 day
Team leadership and group project
    management

D50.   DISCOVERY; USA
−; Interact
High school, college
30-35; 10-15 hours
Early American colonization

D51.   DISNEY LOTTO; UK
−; Waddingtons Ltd
2 years old and over
2-4
Colour and shape recognition

D52.   DISPATCHER; USA
−; Avalon-Hill Inc
General
Railway scheduling game

D53. DISUNIA; 1968; USA
–; El Capitan High School, Interact
Junior high, high school, college
20-35; 2-4 hours
Insight into the problems of disunity in the
USA in 1781-1789, played out on
another planet in 2076.

D54. DIVISION; 1968; USA
P. De Kock, D. Yount; Interact
Grade 7, College
16-35; 5-15 days
Insight into problems facing the electorate
during the 1850s, political compromise,
multiple causation, electoral bargaining

D55. DIVISION; USA
–; D. C. Heath Co; Abt Associates Inc
Elementary school
Practice in arithmetic drill

D56. DOCTOR'S DILEMMA; UK
J. Aitchison, T. Taylor; University of
Glasgow
Secondary school
Mathematics and computing in medicine

D57. THE DOLPHIN PROJECT;
1971; UK
J. K. Jones; J. K. Jones
Secondary school
Controversy in town council meeting and
its reporting

D58. DOMESTIC ELECTRICAL
PRODUCTS; UK
P.E. Consulting Group; P.E. Consulting
Groups
College, management
Management techniques, co-organisation,
production management

D59. DOMINOES; USA
P. McKee, M. L. Harrison; Houghton
Mifflin; Abt Associates Inc
Pre-reading
2-4
Practice in associating consonant letter
forms and sounds

D60. DOMI-NOTES I; 1961; USA
–; G. Schirmer
6 years old and over
2-4
Teaches musical notation

D61. DOMI-NOTES II; 1961; USA
–; G. Schirmer
6 years old and over
2-4
Teaches musical notation

D62. DOMINO NUMBER GAME;
USA
–; D. C. Heath Co
Primary school
Practice in use of factors and multiples

D63. DOMINO WAR GAME; UK
D. S. C. Arthur; Greenfields High School
Secondary school
20-60; 4-8 hours
Post World War II struggles in Vietnam and
Rhodesia

D64. DOMINOES I; UK
D. Glynn; Oliver & Boyd
Pre-readers
Teaches reading

D65. DOMINOES II; UK
D. Glynn; Oliver & Boyd
Pre-readers
Teaches reading

D66. DOMINOES III; UK
D. Glynn; Oliver & Boyd
Pre-readers
Teaches reading

D67. DOMINOES IV; UK
D. Glynn; Oliver & Boyd
Pre-readers
Teaches reading

D68. DOVER PATROL; UK
–; H. Gibson
General
Naval war game

D69. DOUBLE CROSS; UK
—; Waddingtons Ltd
7 years old and over
2-4
Teaches the handling of money, number and
the four rules

D70. DOVER PATROL; UK
—; H. P. Gibson
General
Naval tactics

D71. DOWN WITH THE KING;
1970; USA
T. E. Lineham, D. G. Roach; Herder and
Herder
Primary, junior high, high school
8-35; 3 hours +
Demonstrates revolutionary process; teaches
some medieval history

D72. DRAWING A BOUNDARY;
1970; USA
—; Macmillan Co; High School Geography
Project
High school
6-30; 1 day
Insight into use of common traits in a
cultural heritage to distinguish the
culture's region

D73. DRUG ATTACK; USA
—; Dynamic Design Industries
General
Drug smuggling and drug abuse

D74. DRUG ATTACK GAME;
1969; USA
Lockheed Education Systems; Lockheed
Education Systems
Junior High School
3-8
Insight into problems of drug abuse

D75. THE DRUG DEBATE; USA
K. C. Cohen; Academic Games Associates Inc
Upper elementary school, adult
Up to 35
Insight into conflicts in a controversial area
and the nature of attitude change

D76. THE DUEL; 1878; UK
P. Colomb;
Naval
Insight into gun and torpedo tactics

D77. DUNKERQUE, 1940; USA
—; Simulations Design Corp
General

D78. DUNKIRK; USA
G. Gygax, D. Lowry, C. Johnson; Guidon
Games
General
1940 campaign in France

D79. DUQUESNE UNIVERSITY
MANAGEMENT GAME; USA
Duquesne University; School of Business
Administration, Duquesne University
College, graduate school
30-99; 12 hours
Experience of communication and decision-
making

D80. DYNAMIC BUSINESS GAME;
1972; UK
—; Imperial College
Post graduates
Gives experience in decision-making in
in-company and inter-company co-
operation and competition

D81. THE DYNASTY GAME;
1969; USA
P. C. Huang; Dynasty International Inc
Junior high, high school
4-32; 2¾ hours
Insight into problems of an agricultural
society: population, social welfare,
taxation, revolution, law and order, cost
of public works

E1. EAST FRONT; USA
—; Simulations Publications Inc
General
Second World War

E2. ECOLOGY; 1971; USA
—; Urban Systems Inc
10 years old, adult
Insight into conflict between man and his
environment

E3. ECOLOGY KIT 1: WHY ARE
     LEAVES GREEN; 1971; USA
–; Urban Systems Inc
Primary, junior high, high school
1-3
Teaches the physiology of photosynthesis

E4. ECOLOGY KIT 2: LIFE IN
     THE WATER; 1971; USA
–; Urban Systems Inc
Primary, junior high, high school
1-3
Nutrition and survival of plankton as
  essential food sources for aquatic life

E5. ECOLOGY KIT 3: PREDATOR
     PREY; USA
–; Urban Systems Inc
Primary, junior high, high school
1-3
Survival, food chains, food webs and
  ecological balance

E6. ECOLOGY KIT 4: WHAT
     MOVES LIFE; 1971; USA
Urban Systems Inc
Primary, junior high, high school
1-3
Teaches the relationship of living things to
  the physical factors in their environment

E7. ECOLOGY KIT 5: LIFE FROM
     DEATH; 1971; USA
–; Urban Systems Inc
Primary, junior high, high school
1-3
Contribution of waste to the survival of
  life through recycling of nutrients
  throughout the food web

E8. ECON; USA
–; Center for Environmental Quality
  Management
University
Interaction of economic, demographic and
  social welfare factors in a developing
  nation

E9. ECONOMIC STRATEGY
     ANALYSIS GAME; 1969; USA
A. Packer, J. Ziverneman, J. Wright;
  Research Triangle Institute
College
1-3

E10. ECONOMIC SYSTEM; 1969;
      USA
T. R. Harris, J. Coleman; Western Publishing
  Co; Didactic Systems Inc; Academic Games
  Associates Inc; Johns Hopkins University
Junior high, high school, college
7-13; 2-4 hours
Illustrates the principle of diminishing
  marginal utility

E11. ECONOMY; USA
–; Abt Associates Inc; University of
  Chicago
6th grade
Insight into cyclic flow of goods and
  services in the economic system

E12. ECOPLANY. 1971; France
French National Scientific Centre; Editions
  Ouvrieres
Adults
Insight into economic problems

E13. ECOPOLIS; USA
–; Interact
High school, college
30-35; 10-15 hours
Problems of ecological history and
  contempory environment

E14. EDGE CITY COLLEGE;
      1971; USA
L. A. Callahan et al.; Urbandyne
High school, college
20-30
Overview of the functioning of a college

E15. THE EDISON ELEMENTARY
      PRINCIPALSHIP; 1966; USA
T. Atkins; University Council for Educational
  Administration
Education administrators
1-25
Experience of an elementary school
  principalship

E16.   ED PLAN; USA
C. Abt, R. Glazier, P. S. Miller,
    E. Gottheil; Games Central
15 years old to adult
29-36
Major educational planning issues; insight
    into alternative plans, costs and benefits

E17.   EDSIM; 1968; UK
−; Institute of Development Studies
Adults
2 days
Therapy for intellectual cramp through
    economics

E18.   EDUCATION GAME; 1968; UK
−; Mother
General
1; 5 minutes
Insight into education system

E19.   EDUCATIONAL SYSTEM
        PLANNING GAME; USA
−; Abt Associates Inc
Educators, students
2+
Educational and innovative planning in
    educational policy

E20.   THE EGG AND EVOLUTION;
        USA
T. E. Lineham, M. L. Prendergast; Herder
    and Herder
High School
Insight into the processes of organic
    evolution

E21.   EI PRACTICE; 1962; USA
L. E. Allen; Wff'n Proof Inc
Grade 1 to college
2+
Teaches use of equivalence−in rule of logic

E22.   1812; USA
−; Simulations Publications Inc
General
Napoleonic campaigns

E23.   EINSTEIN GAMES; 1973; UK
R. E. M. Shaw; Times Educational
    Supplement
General
1
Teaches facts of physics

E24.   ELECTION; 1968; USA
J. Young; Educational Games Co
Primary, junior high, high school, college
4; 30-60 minutes
Insight into the fundamentals of the
    democratic process and the elective
    system

E25.   ELECTION CAMPAIGN GAME;
        UK
J. P. Cole
−; Ideas in Geography
Secondary school
2-20; 1-12 hours
Presidential election (USA)

E26.   AN ELECTION CAMPAIGN
        PLANNING GAME; 1971; UK
J. P. Cole; Geography Department,
    University of Nottingham
Secondary school
1
Insight into the planning problems of an
    election campaign

E27.   ELEMENTARY PRINCIPAL-
        SHIP GAME 1: PARENT/
        TEACHER CONFLICT; 1969;
        USA
R. Ohm, T. W. Wiggins; University Council
    for Educational Administration
Educational Administrators
Conflict resolution

E28.   ELEMENTARY PRINCIPAL-
        SHIP GAME 2: TEACHER
        CONFLICT; 1969; USA
R. Ohm, T. W. Wiggin; University Council
    for Educational Administration
Educational administrators
Conflict resolution

E29.   EL SOMBRERO CAFÉ; UK
—; Berkshire College of Education
Secondary school
Teenage behaviour in a cafe; drug-pushing

E30.   EMPEROR OF CHINA; USA
—; Dynamic Design Industries
General
Unification of China

E31.   EMPIRE; 1964; USA
R. Branfon, C. Abt; KDI Instructional
   Systems Inc; Abt Associates Inc;
   Education Development Center
Grade 8, junior high school
15-35; 45 minutes
Insight into mercantilism, set in 18th
   century immediately after independence
   of USA

E32.   THE EMPLOYMENT
       MARKET PLACE; 1969; USA
J. LeVaux, R. E. Horn; Information
   Resources Inc
High school
10-15; 4-8 hours
Insight into both sides of job-hunting/staff-
   finding; training in job interviews

E33.   ENGLISH LOCAL
       GOVERNMENT; 1970; UK
R. H. R. Armstrong, M. Hobson; Institute
   of Local Government Studies,
   Birmingham University
Post-experience, post-graduate, local
   government
4+; 2 hours +
Application of NEXUS (q.v.) to study of
   local government

E34.   THE ENVIRONMENT GAME;
       1967; UK
N. Calder; Secker and Warburg
General
Insight into needs of society in the
   twenty-first century

E35.   ENVIRONMENTAL
       MANAGEMENT GAME;
       1969; UK
A. Lassiere, G. Hoinville; Ministry of
   Transport (Department of the
   Environment)
Adults

E36.   EO PRACTICE; 1962; USA
L. E. Allen; Wff'n Proof Inc
Grade 1 to college
2+
Use of equivalence-out rule of logic

E37.   EQUALITY; USA
—; Interact
High school, college
30-35; 10-15 hours
Racial struggle in an American city

E38.   EQUALS; USA
—; D. C. Heath Co; Abt Associates Inc
Elementary school
Practice in arithmetic drills

E39.   EQUATION MATCH; USA
—; D. C. Heath Co; Abt Associates Inc
Elementary school
Practice in arithmetic drills

E40.   EQUATION RUMMY; USA
—; D. C. Heath Co; Abt Associates Inc
Elementary school
Practice in arithmetic drills

E41.   EQUATIONS; 1969; USA
L. A. Allen; Wff'n Proof Inc
Grades 4-12
2+
Addition, subtraction, multiplication,
   division, exponentiation, radicals, use of
   different number bases

E42.   EQUIPMENT EVALUATION;
       1968; USA
E. Rausch; Didactic Systems Inc; Science
   Research Associates
Junior, middle management
6+; 2 hours
Decision-making on new capital expenditure

E43. ESCALADO; UK
—; Chad Valley Ltd
General
Horse racing

E44. THE ESSENTIALS OF BOYS'
CLUB LEADERSHIP; UK
H. E. Higgins; National Association of Boys
Clubs
Youth leaders
Club leadership 'in-tray' practice

E45. ESSO MANAGEMENT
GAME; UK
S. Hargreaves; Esso Petroleum Co
Management
Planning a consumer durable product

E46. ESSO REFINERY
SIMULATION GAME; UK
—; Esso Petroleum Co
Management

E47. ESSO RESEARCH AND
ENGINEERING ALCOHOL
PLANT MODEL; UK
—; Esso Petroleum Co
Management

E48. ESSO SERVICE STATION
EXERCISE; UK
—; Esso Petroleum Co
Management
Insight into the problems of competition
and changing patterns of demand

E49. ESSO STUDENTS BUSINESS
GAME; UK
—; Esso Petroleum Co
College, management
Insight into interaction of decisions, team
nature of management and use of
accounting procedures

E50. ESTABLISHING JOINT
CONSULTATION; 1966; UK
J. Paterson; Nelson Ltd
Management
Human problems at all levels

E51. ... ET ALIA; 1972; UK
R. H. R. Armstrong, M. Hobson, J. R. Hunter;
Institute of Local Government Studies,
Birmingham University
Undergraduate upwards
6+; 1½-2 hours
A game with abstract content designed to
develop an understanding of aspects of
the manner in which groups decide on
objectives

E52. EUROCARD; 1967; USA
R. Allen, J. M. Clark; International Learning
Corp.; Nova Learning Corp.
Pre-school, primary, junior high, high
school
2-4; 30 minutes
Geography of Europe; book searches and
the effective organisation and précis of
information

E53. EVADE; USA
—; 3M Company
General
2
Tactical board game

E54. EVES GOOD FOODS; UK
P.E. Consulting Group; P.E. Consulting
Group
College, management
1+
A business appraisal game, particularly in
retail distribution

E55. EXECUTIVE ACTION
SIMULATION; 1960; USA
L. W. Herron; Prentice-Hall Inc
Management
2-12
Practice in decision-making; insight into
relationships between functions and the
neccesity for decisions based on
insufficient data

E56. EXECUTIVE CONFERENCE;
1968; USA
A. A. Zoll; Addison-Wesley
Management
1
Self-assessment of strengths and weaknesses

E57.   EXECUTIVE DECISION; USA
S. Sackson; 3M Company
General
2-6; 1 hour
Players purchase raw materials, manufacture
  products and attempt to sell at the best
  market price

E58   THE EXECUTIVE GAME;
  1966; USA
R. C. Henshaw, J. R. Jackson; Richard
  D. Irwin Inc
College, graduate school, management
9; 1-60 days
Practice in forward planning and decision-
  making

E59.   EXECUTIVE MANAGEMENT
  EXERCISE; 1965; UK
A. Life, B. Eldon, J. N. Fairhead, D. S. Pugh,
  W. J. Williams; E.U.P. Ltd
Management
Insight into policy-making, co-ordination and
  control

E60.   EXECUTIVE SIMULATION
  GAME; USA
W. D. Heier; College of Business
  Administration, University of Arizona
College, management
16-35; 2 hours
Consequences of poor planning; need for
  systematic rather than ad-hoc planning

E61.   EXERCISE ATTITUDES; USA
B. M. Bass; Didactic Systems Inc; Instad
Management trainees
4-12
Insight into expectations about work
  satisfaction as material gain, interest,
  leadership, policies and procedures

E62.   EXERCISE COMMUNICATION;
  USA
B. M. Bass; Didactic Systems Inc; Instad
Management trainees
4-12
Exploration of one and two-way
  communication

E63.   EXERCISE COMPENSATION;
  USA
B. M. Bass; Didactic Systems Inc; Instad
Management trainees
4-12
Insight into merits of performance valuing
  and effects on job performance;
  difficulty of having to live with an
  unpopular decision

E64.   EXERCISE EVALUATION; USA
B. M. Bass; Didactic Systems Inc; Instad
Management trainees
4-12
Expresses feelings about training and allows
  comparison of intuitive and systematic
  approaches to evaluating decisions

E65.   EXERCISE FISHBOWL; USA
B. M. Bass; Didactic Systems Inc; Instad
Management trainees
4-12
Opportunity to evaluate group behaviour on
  the basis of trust, decision process, goal
  clarity, resource utilisation

E66.   EXCERCISE FUTURE; USA
Miller, Haas, Bass and Ryterband; Didactic
  Systems Inc; Instad
Management trainees
4-12
Insight into future work expectations and
  implications of training for trainee

E67.   EXERCISE IN HUMAN
  RELATIONS; UK
R. W. Elliot; National Association of Youth
  Clubs
Teenagers and upwards
Insight into misunderstandings

E68.   EXCERCISE KOLOMON; USA
Thiagarajan, Bass, Ryterband; Didactic
  Systems Inc; Instad
Management trainees
4-12
Exploration of willingness to take risks as
  decision-makers (individual or group);
  relative importance of pay-off, possibility
  of success and time.

E69. EXERCISE LEMMING; 1968;
UK
N. Rints, D. Hayes; Department of
Management, Slough College of Technology
Management students
Problems of production, marketing and
finance in a firm in a competitive
situation

E70. EXERCISE LIFE GOALS;
USA
B. M. Bass; Didactic Systems Inc; Instad
Management trainees
4-12
Estimation of own life-goals and work
attitudes

E71. EXERCISE NEGOTIATIONS;
USA
B. M. Bass; Didactic Systems Inc; Instad
Management trainees
4-12
Shows how the relationship of commit-
ment to groups' strategy can seriously
limit a negotiator's behaviour

E72. EXERCISE OBJECTIVES; USA
B. M. Bass; Didactic Systems Inc; Instad
Management trainees
4-12
Relationship of perceived organizational
goals and effects of perceptions on
decision making demonstrates that a
manager is a 'systems balancer'

E73. EXERCISE ON THE
CONTROL FUNCTION; USA
A. A. Zoll; Addison-Wesley
Management
1
Control problems and solutions

E74. EXERCISE ORGANISATION;
USA
B. M. Bass; Didactic Systems Inc; Instad
Management trainees
4-12
Insight into problems of developing and
carrying out plans in large organisations
and into intergroup conflict

E75. EXERCISE QUINTAIN; UK
–; John Laing Ltd
Management
Economics of building and civil engineering
firms

E76. EXERCISE SELF APPRAISAL;
USA
B. M. Bass; Didactic Systems Inc; Instad
Management trainees
4-12
Insight into own behaviour and learning
and relations with others; insight into
management

E77. EXERCISE SUCCESS; USA
B. M. Bass, S. D. Deep; Didactic Systems Inc;
Instad
Management trainees
4-12
Insight into the factors which lead to success
in an organisation

E78 EXERCISE SUPERVISE; USA
B. M. Bass; Didactic Systems Inc; Instad
Management trainees
4-12
Demonstrates leadership and subordinate
styles in a task situation

E79. EXERCISE VENTURE; USA
Link, Thiagarajan, Trbovich, Vaughan;
Didactic Systems Inc.; Instad
Management trainees
4-12
Insight into individual and group planning
of corporate strategy

E80. EXERCISE WILLING HORSE;
1968; UK
I. McHorton; Senior Secretaries Ltd
Junior management
12-25; 1 day
Crisis situation simulation on problems in
supplying an order

E81.　EXLAND; 1970; UK
R. H. R. Armstrong, M. Hobson; Institute
of Local Government Studies, Birmingham
University
Post-experience, post-graduate, local
government
4+; 2 hours +
Application of NEXUS (q.v.) to study of
political and economic problems of a
developing country

E82.　EXMARK; 1969; UK
C. Loveluck; Management Games Ltd
6th form and upwards, management
Up to 24; 9 hours minimum
Insight into competetive export marketing;
decisions on sales forecasting production,
product modification, finance, budgeting
and stock control, price fixing, sales
force activity, agents, sales promotions

E83.　EXPLANATION; USA
C. H. Adair, R. F. Allen; Teacher Research;
Simulation Exercises and Games
Junior high, high school, college,
management
6
Formulation of questions, explanations of
answers; theories explaining major social
problems

E84.　EXPLORATION; UK
J. C. Spiring; Waddingtons Ltd
8 years old and over
2-4
Preparing and mounting of expeditions

E85.　EXPLORERS I; USA
–; Project Simile
High school
18-35; 4-5 hours
Early exploration of North America

E86.　EXPLORERS II; USA
–; Project Simile
High school
Exploration of South America

E87.　THE EXPORT DRIVE GAME;
1969; UK
R. Walford; Longmans Green
Secondary school
Insight into problems of exporters and
U.K's main exports and markets

E88.　EXTINCTION: THE GAME
OF ECOLOGY; 1971; USA
–; Sinauer Associates, Stamford Corp
High school
1-2 hours
Insight into ecological processes and chain
reactions to natural and man-made changes

E89.　EXTRACOL; 1969; UK
Cooke; Ministry of Public Building and
Works (Department of the Environment)
Management
Problems of co-operation in a building team

F1.　FACTOPLAN; 1967; UK
–; Polycon Building Industry Consultants
Management
Problems of co-operation in a building team

F2.　FACTORY; 1971; USA
C. H. Kriebal; Carnegie-Mellon University
Advanced management students
Exploratory analysis in sales forecasting,
aggregate planning, production smooth-
ing, inventory control, management
information reporting

F3.　FACTS IN FIVE; USA
–; 3M Company
General
Any number; 1 hour
Vocabulary, word building, general
knowledge

F4.　FACULTY CARDS; 1963; USA
–; ACO Games Division, Allen Co Inc
General
2-4; ½-3 hours
Builds vocabulary, teaches phonics and
syllables

F5. FAMILY BUDGET; 1964; USA
—; Creative Studies Inc
Junior High School
16-40; 2-4 hours
Practice in quantitative skills, estimating,
computing fractions, rounding off sums
of money to nearest dollar

F6. THE FAMILY GAME; USA
—; Psychology Today
High school
Insight into roles, including own, within
family

F7. FAMILY PAIRS; UK
E. Root; Rupert Hart-Davis
Pre-readers
Teaches consonant blends

F8. FARM MANAGEMENT
SIMULATION; 1968; USA
F. Smith, S. D. Miles, Oregon State
University
Management
4-30; 3-6 hours
Economics of farm management

F9. FARMING; USA
—; Collier Macmillan Inc
High school
15-30; 40-50 hours
Insight into agriculture

F10. THE FASHION GAME; UK
—; National Computing Centre
Secondary school
Reinforces ideas of information processing
and computers

F11. FAT CAT; USA
P. Lamb; Lyons & Carnahan
Grades 1-6
3-8
Emphasizes reinforcement experience with
beginning and ending consonants

F12. THE FATE OF OUR FOOD;
1969; USA
B. Belkin; Creative Studies Inc
Primary, junior high school
2-6; 25-90 minutes
Teaches physiology of digestion

F13. FEDERAL MARKETPLACE;
1968; USA
R. E. Horn; Information Resources Inc
Management, political science and
education students
15+; 8 hours
Insight into strategy of seeking research
funds

F14. FEDERAL RESERVE
SYSTEM GAME; 1966; USA
A. A. Zoll; Addison-Wesley
Adults
Insight into banking operations

F15. FEUDAL; USA
—; 3M Company
General
2-6; 1 hour
Strategic board game on medieval warfare

F16. FIDUCIARY ACTIVITY
SIMULATION TRAINING
(FAST); 1966; USA
Abt Associates Inc; United States Trust Co
Management
6-12; 1-2 days
Practice in trust and investment
administration

F17. FIGHT IN THE SKIES; USA
—; Guidon Games
General

F18. FIGHTING SAIL; 1965; UK
E. Solomon; Games & Puzzles
General
2
Insight into 19th century fleet warfare

F19. FINANSIM: A FINANCIAL
MANAGEMENT SIMULATION;
1967; USA
P. S. Greenlaw, W. Frey; Didactic Systems
Inc; International Textbook Co
College
2+; 2 hours
Insight into financial management problems

F20.   FIND IT LOTTO; 1972; UK
A Stephenson; Galt Toys
4-6 years old
Improves observation

F21.   FIND OUT WHAT IT FEELS
       LIKE TO BE A COMPUTER:
       GAME I; UK
S. Gill; Software Sciences Ltd
General
1
Insight into multiplication of two numbers
   X and Y whose answer is 2

F22.   FIND OUT WHAT IT FEELS
       LIKE TO BE A COMPUTER:
       GAME II; UK
S. Gill; Software Sciences Ltd
General
1
Finding highest common factor and lowest
   common multiple of two numbers

F23.   FIND OUT WHAT IT FEELS
       LIKE TO BE A COMPUTER:
       GAME III; UK
S. Gill; Software Sciences Ltd
General
1
Division of two whole numbers

F24.   FIND OUT WHAT IT FEELS
       LIKE TO BE A COMPUTER:
       GAME IV; UK
S. Gill; Software Sciences Ltd
General
1
Finding square root of a positive number

F25.   FINDING A MATE; USA
−; Edcom Systems Inc; Abt Associates Inc
Grade 4
Insight into the fact that complex family
   systems are uniquely human and that
   Aranda social patterns are far from
   primitive

F26.   FINMARK; UK
C. N. Aydon; Management Games Ltd
6th form, college, university, management
1-15; 7-8 hours
Practice in financial and simple marketing
   problems in a manufacturing business;
   decisions on pricing, marginal costing,
   break-even analysis, optimum use of
   plant, sub-contracting, cash budgeting,
   sources of finance

F27.   FIREFIGHT; 1969; USA
C. S. Bauer, W. A. Brown; Martin-Marietta
   Corp
Military
4+
Measuring expected weapon system
   performance; illustrates tactical ground
   combat operations

F28.   THE FIRM; 1968; USA
E. Rausch, H. J. Cranmer; Science Research
   Associates, Didactic Systems Inc
High school, college
6+; 2-5 lessons
Teaches the economic basis of the firm,
   fundamental relationships between
   assets, liabilities, revenue, cost, profit
   and net worth

F29.   FIRST ARITHMETIC GAME;
       1957; USA
E. W. Dolch; Garrard Publishing Co
Kindergarten, grade 1
2-4; 15-20 minutes
Creates interest in counting objects and
   develops number awareness

F30.   THE FISHING GAME;
       1958; UK
L. W. Downes, D. Paling; Oxford University
   Press
Primary school
2+
Practice in addition

F31.   THE FIVE GAME; 1967; USA
H. A. Springle; Science Research Associates
Pre-school age to grade 1
1-4; 20-30 minutes
Insight into the nature of equivalence

F32.   FIZZOG; 1972; UK
K. Garland; Galt Toys
5-7 years old
Practice in observation and selection

F33.   THE FLIGHT OF THE
       GOEBEN; USA
—; Simulations Publications Inc
General
Naval warfare in 1914 in Mediterranean

F34.   FLIP (FAMILY LIFE
       INCOME PATTERNS);
       1970; USA
—; Instructional Simulations Inc
Junior high school up to adults
1-40; 1 hour +
Insight into family budgeting, investment,
   credit and interest in terms of changing
   family goals

F35.   FLUTTER; UK
—; Spear's Games Ltd
General
2-8
Insight into stock market operation

F36.   FLYING CIRCUS; USA
—; Simulations Publications Inc
General
½-1 hour
First World War, aerial combat

F37.   FLYING FORTRESS I; USA
—; Simulations Publications Inc
General

F38.   FLYING FORTRESS II; USA
—; Simulations Publications Inc
General

F39.   FLYING TIGERS; USA
—; Simulations Publications Inc
General

F40.   FOIL; USA
—; 3M Company
General
2-4; 1 hour
Vocabulary practice

F41.   FOOD 1; 1969; USA
H. A. Springle; Science Research Associates
Pre-school to grade 1
1-4; 20-30 minutes

F42.   FOOD 2; 1969; USA
H. A. Springle; Science Research Associates
Pre-school to grade 1
1-4; 20-30 minutes

F43.   FOOTBALL EXPRESS; UK
—; Subbuteo Ltd
General
Five-a-side football

F44.   FOOTBALL GAME; 1971; UK
—; Waddingtons Ltd
6 years old and over
2-4
Teaches rules and positions of soccer

F45.   FOOTBALL STRATEGY; USA
—; Avalon-Hill Inc
General
2 or 4
Insight into playing strategies of quarterback
   in American football

F46.   FORTRESS HOLLAND; USA
H. Radice; Guidon Games
General
Insight into the Dunkirk campaign

F47.   THE FOUNDATION GAME
D. Miller; Albion
General
8
A 'Diplomacy' variant based on Asimov's
   science fiction 'Foundation' trilogy

F48.   THE FOUR GAME; 1967; USA
H. A. Springle; Science Research Associates
Pre-school to grade 1
1-4; 20-30 minutes
Insight into nature of equivalence

F49.   FOUR PEGS; UK
—; Galt Toys
Primary school
Insight into space relations

F50. FOUR STRUCTURAL
PROBLEMS IN
ORGANISATION; 1966; UK
C. Margerison; Nelson Ltd
Management
Problems of conflict, organisational change
and social control

F51. FOREIGN POLICY DECISION-
MAKING EXERCISE; USA
W. D. Coplin; Department of Political
Science, Syracuse University
College
5-45; 1-2 hours
Insight into the intellectual, ideological and
organizational factors which effect
decision-making

F52. FORMULA 1; UK
-; Waddingtons Ltd
10 years old and over
2-6
Motor-racing

F53. FOREST SERVICE FIRE
CONTROL SIMULATOR; USA
US Department of Agriculture; US Department Agriculture
Fire control trainees
1-5; $\frac{1}{4}$-4 hours
Teaches principles and techniques of forest
and grassland fire suppression practice in
decision-making

F54. FORMULON; 1971; UK
J. S. Stoane; Bell Baxter Senior High School
Secondary school (3rd year Scotland, 4th
year England)
1-8; 10-15 minutes
Re-inforcement of the teaching of chemical
formulae

F55. FORST: A COMPUTER
SIMULATION OF A
LUMBER COMPANY; Canada
J. D. Burnett; Division of Education,
Research Services, University of Alberta
Junior high, high school
1+; 40-120 minutes
Insight into variables operating in a primary
industry and their interaction and the
necessity for constant evaluation and
modification

F56. 4000 AD; UK
-; Waddingtons Ltd
General
2-4
Interstellar conflict strategy game

F57. 4. 2. 4.; UK
-; Games and Puzzles
8 years old over
Teaches rules of Association football

F58. FRACTONIMOES; 1971; UK
C. M. Glover, A. D. Mackie; Maths Learning
Systems
Primary school
Demonstrates and gives practice in relating
fractions to wholes and to each other

F59. FRANCE 1940; USA
-; Avalon-Hill Inc
General

F60. FRANCO-PRUSSIAN WAR;
USA
-; Simulations Publications Inc
General
Strategy and tactics of 19th century conflict

F61. FRANTIC WFF N PROOF;
1962; USA
L. E. Allen; Wff'n Proof Inc
Grade 1 to college
2+
Further practice in deduction

F62.   FREE ENTERPRISE GAME;
1965; USA
J. M. Leonard; Board of Cooperative
Education Services
Grade 6
1; 37-74 hours
Demonstration of everyday problems of
retailers, the law of supply and demand,
law of diminishing returns and policy
for specialisation, hiring, firing and
wages

F63.   FRENCH HOLIDAY TRAVEL
GAME; 1973; UK
SGS Associates; Longman
Secondary school
Up to 30; 40 minutes +
Motoring tour through France teaching
language, social background, geography
etc.

F64.   FRONTIER; 1970; USA
A. Kaplan; Science Research Associates;
Abt Associates Inc
High school, junior college
Economic and political development of
American North West and South West
1815-1830

F65.   FRONT PAGE; 1972; UK
J. K. Jones; BBC
Secondary school
Insight into sub-editor's job of writing
headlines and meeting deadlines

F66.   FULL HOUSE; USA
A. W. Heilman, R. Helmkamp, A. E. Thomas,
C. J. Corsello; Lyons and Carnahan
Grades 1 to 3
2+
Recognition of sounds and symbols of
vowels and vowel digraphs and dipthongs

F67.   FUNCTION; 1972; UK
C. Elgood; Management Games Ltd
Management, supervisors
Teams of 4 or 6; 2½ hours minimum
Insight into relationship between purchasing,
production, sales, credit control and
accounting

F68.   FUNCTIONS; USA
D. E. Tritten; North Idaho Junior College
High school
2; 1 hour
Practice in construction of complex
algebraic equations

F69.   FUN WITH SUMS: FIRST
STEPS IN COUNTING; UK
−; Spears Games
Primary school
Simple arithmetic practice

F70.   FURNITURE 1; 1969; USA
H. A. Springle; Science Research Associates
Pre-school to grade 1
1-4; 20-30 minutes

F71.   FURNITURE 2; 1969; USA
H. A. Springle; Science Research Associates
Pre-school to grade 1
1-4; 20-30 minutes

G1.   GALAPAGOS; USA
−; Abt Associates Inc
High school
3+
Practice in scientific observation, evaluation,
prediction and theory formation; set in
the Galapagos Islands and based on the
evolution of fishes

G2   GAME; 1971; USA
−; Learning Institute of North Carolina
Educators
39-40; 3 hours
Insight into interdependence of educational
programmes and their environments

G3.   THE GAME; 1971; UK
A. Ighodaro, M. Ford; Think Games Ltd
Adults
2

G4.   GAME 1; 1969; UK
J. Rae; Hornsey College of Art
College
6; 1 hour
Insight into group behaviour and problem-
solving

G5. GAME 6; 1969; UK
J. Rae; Hornsey College of Art
College
20-30; ½ day
Insight into communications patterns

G6. GAME 7; 1969; UK
J. Rae; Hornsey College of Art
College
20-30; 1 day
Insight into roles, communication, group
dynamics and problem-solving

G7. GAME CITY; 1969; USA
M. Clayton, R. Goetze, C. Field, R. Hollister,
J. Hester; WGBH Educational Foundation
High school
15+; 2-6 hours
Demonstrates how government solves
problems by identifying consensus
among divergent views

**G8. THE GAME OF THE CLANS
(SCOTTOMACY);**
W. Hoheisel; Albion
General
9
A 'Diplomacy' variant, England versus
Scottish clans

**G9. GAME MODEL FOR INITIAL
CLASS PERIODS OF ORGAN-
ISATION AND ADMINIS-
TRATION OF HEALTH AND
PHYSICAL EDUCATION; USA**
K. G. Tillman; Trenton State College
College
7+; 50 minutes
Establishes classroom rapport and teaches
definition of physical education

**G10. THE GAME OF ANARCHY;
USA**
D. Miller; Albion
General
Diplomacy variant

**G11. THE GAME OF
ANONYMITY; USA**
D. Miller; Albion
General
Diplomacy variant

G12. THE GAME OF CHAOS; USA
D. Miller; Albion
General
Diplomacy variant

**G13. THE GAME OF FARMING;
1969; USA**
–; MacMillan Co; High School Geography
Project
High school
15-30; 4½-6 days
Practice in decision making in areas
affected by chance and natural and
economic factors

**G14. THE GAME OF NATIONS
1973; UK**
M. Copeland; Waddingtons Ltd
Adults and older children
2-4
Political strategy game

G15. GAME OF GOOSE; Holland
–; Jumbo Games
General
Insight into man's life process and life span

G16. THE GAME OF LIFE; UK
J. Conway; J. Conway
Investigation of growth and decay patterns

**G17. THE GAMES PEOPLE PLAY;
1967; USA**
E. Berne, *et al.*; MASCO
Junior high, high school, college
1-8
Develops skill in recognizing social actions
and their consequences

**G18. A GAMING-SIMULATION
EXERCISE ON THE
THEORETICAL
IMPLEMENTATION OF
THE KERNER COMMISSION
REPORT; 1971; Canada**
I. Armillas; Nova Scotia Technical College
College
26; 10 hours
Insight into the reports proposals and
consequences and evaluation of them

G19. A GENERAL AGRICULTURAL FIRM SIMULATOR; 1968; USA
R. Hutton, H. Hinman; Department of Agricultural Economics, Pennsylvania State University
High school, college, graduate school
1+; ½ hour +
Practice in experience of management policy making and its applications

G20. GENERAL ELECTRIC MANAGEMENT GAME 1; 1969; USA
R. W. Newman; General Electric Co
Management
1-10; ¼-25 hours
Learning to manage other people; strategic thinking

G21. GENERAL ELECTRIC MANAGEMENT GAME 2; 1969; USA
R. W. Newman; General Electric Co
Management
2-14; 30 minutes
Practice in tactical planning in sales

G22. GENERAL ELECTRIC MANAGEMENT GAME 3; 1969; USA
R. W. Newman; General Electric Co
Management
2-14; ½ hour
Planning and managing a business and its personnel

G23. GENERAL ELECTRIC MANAGEMENT GAME 4; 1969; USA
R. W. Newman; General Electric Co
Management
2-14; ½ hour
Helps personnel relations staff to take a systems view of their job

G24. GENERAL MANAGEMENT IN-BASKET; USA
A. A. Zoll; Addison-Wesley
Management
1

G25. GENERAL MANAGEMENT SIMULATION; USA
—; R.C.A.
Management

G26. GENERAL MANAGEMENT SIMULATION; Japan
—; Tokyo Center for Economic Research
Management

G27. GENERATION GAP; 1969; USA
E. O. Schild, S. S. Boocock; Western Publishing Co; Academic Games Associates
Junior high, high school
4-6; ½-1 hour
Insight into structure of power and strategies for handling the conflict it generates

G28. GENEX; 1965; USA
L. Bloomfield; Massachusetts Institute of Technology
Management
Insight into business crisis situations

G29. GERMANY VERSUS THE WORLD; UK
F. C. Davis; Albion
General
7
A 'Diplomacy' variant, based on World War Two

G30. GERRYMANDER; 1971; UK
J. P. Cole; Department of Geography, Nottingham University
Secondary school upwards
Rigging the electoral boundaries in a city to ensure a majority for a particular party

G31. GET SET: GAMES FOR BEGINNING READERS; 1967; USA
P. McKee, M. L. Harrison; Houghton-Mifflin
Pre-reading;
Practice in skills of left-to-right sequence, discrimation of consonant letter sounds and forms and supplying missing words from context

G32.   GET SET: THE NUMBER
        GAME; 1971; UK
C. M. Glover, A. D. Mackie; Maths Learning
    Systems
Primary school
2-4
Teaches concept of set and Venn diagrams

G33.   GET SET: THE PEOPLE
        GAME; 1971; UK
C. M. Glover, A. D. Mackie; Maths Learning
    Systems
Primary school
2-4
Teaches concept of set and venn diagrams

G34.   GETTYSBERG; 1964; USA
—; Avalon-Hill Inc
General
2+; 2-4 hours
Insight into Robert E. Lee's campaign in
    American Civil War; practice of strategic
    and predictive thinking

G35.   GETTYSBERG; USA
—; Simulations Publications Inc
General

G36.   GHETTO; 1969; USA
D. Toll; Western Publishing Co; Academic
    Games Associates
Junior high school to adults
7-10; 2-4 hours
Insight into pressures on urban poor and the
    decisions they have to make

G37.   GHQ; UK
—; Waddingtons Ltd
General
War game set in 1930s Europe

G38.   GITHAKA; USA
—; Edcom Systems Inc; Abt Associates Inc
Grade 4
Illustrates problems of founding a community,
    including acquisition of land and wives, in a
    rural district of central Kenya

G39.   GLAD LAD; USA
P. Lamb; Lyons and Carnahan
Grades 1-6
3-6
Practice in making beginning and final
    consonant substitutions

G40.   GO; 1968; USA
—; Simulations Publications Inc
General

G41.   GO; UK
—; Waddingtons Ltd
10 years old and over
2-6
A travel game involving the planning of
    routes (time taken, costs, etc.)

G42.   GO: A GAME OF STRATEGY;
        1971; USA
—; Urbandyne
General
2
Practice in strategic thinking

G43.   GO FOR BROKE; UK
—; Invicta
General
Financial management

G44.   GOALS FOR DALLAS;
        1970; UK
R. H. R. Armstrong, M. Hobson; Institute of
    Local Government Studies, Birmingham
    University
Post-experience, post-graduate
4+; 2 hours +
An application of NEXUS (q.v.) to city
    planning

G45.   GODFATHER GAME; 1972;
        USA
—; Family Games Inc
General
Mafia operations

G46.   GOING TO SCHOOL BY
        'BUS'; 1971; UK
—; Mathematical Pie
Maths students
1
Teaches the idea of a flow diagram

G47. GOLDBERG V. SILVER-
STEIN; 1973; UK
E. Solomon; Games & Puzzles
General
2
Insight into business world in clothing
industry

G48. THE GOOD SOCIETY
EXERCISE; 1971; USA
W. D. Coplin, S. Alpert; International
Relations Programme, Syracuse University
High school
1+
Insight into justice and order in social
systems

G49. GRAND ARMEE; USA
−; Simulations Publications Inc
General
Napoleonic warfare

G50. GRAND STRATEGY; USA
C. C. Abt, R. Glazier; Games Central
15-18 years old
10-30
Insight into international relations and
diplomacy

G51. THE GREAT DOWNHILL
SKI GAME; USA
−; Franklin Merchandising
General
1-4

G52. THE GREAT GAME OF
BRITAIN; 1973; UK
−; Condor Ltd
General
Board game based on train journey through
Britain

G53. THE GREAT GAME OF
LEGISLATURE; 1963; USA
−; The John Hopkins Magazine
High school
Insight into legislative processes

G54. GREMEX (GODDARD
RESEARCH ENGINEERING
MANAGEMENT EXERCISE);
USA
M. J. Vaccaro; Information Center,
University of Georgia
Management
12; 40 hours
Teaches R and D project management

G55. GRENADIER; USA
−; Simulations Publications
General
Warfare in period 1680 to 1850

G56. GRIEVANCE; UK
−; Guardian Business Services
Management
Insight into the problems of industrial
relationships

G 57. GRIEVANCE HANDLING;
1970; USA
−; Didactic Systems Inc.
College, supervisors, foremen
3+; 2-3 hours
Practice in grievance handling

G58. G.R.I.P.S. 1 (HOLISM); 1971;
UK
R. Levy, R. J. Talbot, L. D. Warshaw;
University of Manchester
University
4-8; 4-5 hours
Introduction to the origins, identification,
structuring and solving of problems

G59. G.R.I.P.S. 2 (ROLE PLAYING);
1971; UK
R. Levy, R. J. Talbot, L. D. Warshaw;
University of Manchester
University
4-8; 4-5 hours
Introduction to the origins, identification,
structuring and solving of problems

G60.   G.R.I.P.S. 3 (PROBLEM
       GENERATING); 1971; UK
L. D. Warshaw, R. Levy, R. J. Talbot;
   University of Manchester
University
4-8; 4-5 hours
Introduction to the origins, identification,
   structuring and solving of problems

G61.   G.R.I.P.S. 4 (PROBLEM
       PICTURING); 1971; UK
L. D. Warshaw, R. Levy, R. J. Talbot;
   University of Manchester
University
1; 4-5 hours
Introduction to the origins, identification,
   structuring and solving of problems

G62.   GROUP SOUNDING GAME;
       1945; USA
E. W. Dolch; Garrard Publishing Co
Grade 2 upwards
2-6; 10-15 minutes
Matching phonics with visual letters and
   groups of letters

G63.   GROUP WORD TEACHING;
       1944; USA
E. W. Dolch; Garrard Publishing Co
Grade 2 upwards
1-6; 10-15 minutes
Teaches word recognition

G64.   GRUNT; USA
–; Simulations Publications
General
Ground combat in Vietnam

G65.   GSPIR MANAGEMENT
       PLANNING DECISION
       EXERCISE; 1966; USA
F. Hendricks; University of Pittsburgh
Management

G66.   GUADAL CANAL; USA
–; Simulations Publications
General
World War II

G67.   GUADAL CANAL; 1966; USA
D. L. Dickson; Avalon-Hill Inc
General
2+; 3-6 hours
Insight into jungle warfare; practice strategic
   and predictive thinking

G68.   GUNS OR BUTTER; USA
W. A. Nesbitt; New York State Education
   Department
High school
20; 1½ hours +
Insight into the processes which lead to
   security and political and economic
   welfare of a state

H1.   HANDICAP GOLF; USA
–; Sports Illustrated Games
General
1-4

H2.   HANDLING CONFLICT IN
      MANAGEMENT: 1 CONFLICT
      AMONG PEERS; 1969; USA
E. Rausch, W. Wohlking; Didactic
   Systems Inc; American Management
   Association
Management students
4+; 2½-3 hours
Conflict resolution

H3.   HANDLING CONFLICT IN
      MANAGEMENT: 2 SUPERIOR/
      SUBORDINATE CONFLICT;
      1969; USA
E. Rausch, W. Wohlking; Didactive Systems
   Inc; American Management Association
Management students
4+; 2½-3 hours
Conflict resolution

H4   HANDLING CONFLICT IN
      MANAGEMENT: 3 SUPERIOR/
      SUBORDINATE GROUP
      CONFLICT; 1969; USA
E. Rausch, W. Wohlking; Didactic Systems
   Inc; American Management Association
Management students
4+; 2½-3 hours
Conflict resolution

H5.   HANG UP; 1969; USA
Synetics Education Systems; Unitarian
  Universalist Association
High school, college, graduate school
3-6; 1 hour +
Promotes empathy and tolerance in order to
  change white racialist attitudes

H6.   HANNIBAL; 1969; USA
−; Simulations Publications Inc
General
Strategy and tactics

H7.   THE HAPPENING GAME;
        1969; USA
−; Community Makers Ltd
Primary, junior high school
1-10; 10-60 minutes
Develops analytical thinking and decision-
  making by sequencing a story

H8.   HAPPY BEARS; 1956; USA
E. W. Dolch; Garrard Publishing Co
Kindergarten, grade 1
2-4; 10-15 minutes
Teaches vocabulary and develops reading
  readiness

H9   HAPPY WORDS; 1971; UK
J. Hicks, T. Kremer; MacDonald
Primary, lower secondary school
Practice in using phonics of initial consonant
  blends

H10.   THE HARD ROCK MINE
          STRIKE; 1970; USA
V. Durham, R. Durham; Random House
Junior high, high school
11-35; 120 minutes
Insight into labour/management conflicts in
  late 19th and early 20th century

H11.   HARVARD BUSINESS
          REVIEW GAME; 1958; USA
G. R. Audlinger, J. R. Greene; Harvard
  University
Management

H12.   HARVARD BUSINESS SCHOOL
          MANAGEMENT SIMULATION;
          USA
−; Harvard University
Management

H13.   HAUNTED HOUSE; UK
−; Denys Fisher Toys
General
2-4
Journey with folk-lore learning

H14.   THE HAYMARKET GAME;
          1969; USA
D. Dal Porto; D. Dal Porto
High school
23-35; 5-6 days
Insight into issues involving labour and
  business in years following industrialisation
  of USA; learning of court procedure;
  examination of question of free speech

H15.   HEADING FOR CHANGE; UK
W. Taylor; Harlech Television; University of
  Bristol
Teachers
Management of innovation in secondary
  schools

H16.   HEALTH EDUCATION
          PLANNING GAME; USA
−; Abt Associates Inc
College, management
Setting objectives in health education for
  hospitals, medical schools and the
  community; developing education and
  allocating funds

H17.   HELMSMAN; UK
−; Triang Ltd
10 years old
2+
Teaches rules of small boat sailing

H18.   HELP; USA
R. Glazier, M. K. Moore, K. Weiner;
  Games Central
8-12 years old
5-32
Practice in decision-making and verbal
  skills; teaches knowledge of
  community emergency agencies

H19.   HEREFORDSHIRE FARM; UK
W. V. Tidswell; Geographical Association
Secondary school
Insight into agriculture

H20.   HEX; 1945; Denmark
P. Hien; Spears Games
4 years old and over
2-6
Teaches visual discrimination and pattern
   recognition

H21.   HEXADIANGLE; 1972; UK
G. Austwick; Mathematics Teaching
2+
Object is to build shapes with sides of equal
   length

H22.   HIDE AND SEEK; UK
—; Rupert Hart-Davis
Primary school
Teaches vocabulary

H23.   HIE SOMA; Denmark
—; Piet Hein
General

H24.   HIGH BID; USA
—; 3M Company
General
2-5; 45 minutes
Auction in which players compete for
   paintings etc.

H25.   HOCUS; UK
R. Hills; P. E. Consulting Group
College, management
Simulation for investigation of queuing
   problems. Can be used manually then
   transferred to computer

H26.   HOMEWORK; UK
P. J. Tansey; Berkshire College of Education
Education students
1
Practice in problem-solving

H27.   THE HOSPITAL EXERCISE
       IN LONG TERM PLANNING;
       UK
—; Wessex Regional Hospital Board
Management
Gives training in hospital planning

H28.   HOMESTEADERS; USA
—; Project Simile
High school
18-35; 5-7 hours
Homesteading in 1870s and 1880s in USA

H29.   HORIZON: A GAME OF WAR
       UK
—; British Broadcasting Corporation
General
1 hour
Arab-Israeli conflict

H30.   HOSPITEX; 1972; UK
H. R. Pollard; Management Games Ltd
Nursing students
6-30; 3-6 hours
Hospital ward management; normal and
   abnormal clinical and ward management,
   decisions have to be made under pressure

H31.   HOUSE PURCHASE; 1970; UK
P. Crean; Manchester College of Education
15 years old and over
Demonstrates role of judgement in balancing,
   budgets, rates, repairs, savings; motivation
   into 'success'

H32.   THE HOUSE THAT JACK
       BUILT; UK
—; Spears Games
4-8 years old
2-4
Colour matching and discrimination practice

H33.   HULL DOWN; 1972; Canada
G. Bradford; G. Bradford
General
Armoured combat

H34.   HUMAN BODY 1; 1969; USA
H. A. Springle; Science Research Associates
Pre-school to grade 1
1-4; 20-30 minutes

H35.   HUMAN BODY 2; 1969; USA
H. A. Springle; Science Research Associates
Pre-school to grade 1
1-4; 20-30 minutes

H36.   HUMAN RELATIONS
        ACTION MAZE; 1966; USA
A. A. Zoll, Addison-Wesley
Management
1
Human relations for the supervisor

H37.   HUMAN RELATIONS: ONE
        DIMENSION OF TEACHING;
        1971; USA
E. G. Buffie; School of Education, Indiana
University
College, education students
12+; 4 hours
Insight into and practice of processes of
    human relations

H38.   HUNTING GAME; USA
—; Edcom Systems Inc; Abt Associates Inc
Grade 4
Survival in a polar region

H39.   HYPERSPACE; USA
A. Calhamer; A. Calhamer
General
Four-dimensional strategy game

H40.   HYPERSPACE DIPLOMACY;
        USA
D. Miller; Albion
Diplomacy variant

I1.   ILDRIDGE GAZETTE; 1971; UK
J. Finlay; Community Service Volunteers
School
24+
Insight into problems of planning a motor-
    way and strategies for solving them

I2.   IMAGINIT MANAGEMENT
        GAME; 1971; USA
R. F. Barton; Texas Technical University;
Allyn & Bacon
College, graduate school, management
4+; 7-20 weeks
Training in functional management and
    integration of functions into whole

I3.   IMDEG (INSURANCE
        MANAGEMENT DECISION
        GAME); 1960; USA
B. Schorr; Travellers Insurance Co
Management
8-40; 8 hours
Practice in analysis, integration, weighing and
    making of decisions about inter-related
    factors

I4.   IMPACT; 1969; USA
R. C. Klietsch; Instructional Simulations Inc
Junior high school to adult
20-40; 4-10 hours
Insight into the way individual and collective
    actions affect a community

I5.   IMPACT: A YOUTH MINISTRY
        SIMULATION; 1970; USA
R. Shukraft, J. Washburn; J. Washburn; San
        Francisco Theological Seminary
Junior high, high school, parents, youth
        advisors, ministers
30-56; 3-24 hours
Testing of various forms of youth ministry

I6.   IMPORT; USA
—; Project Simile
18-35; 6-10 hours
Importing in 6 parts of the world

I7.   THE IN-BASKET EXERCISE;
        1961; USA
Veterans Administration; Veterans
        Administration
Management
8-24; 1½-3 hours
Teaches how to bridge the gap between
        training and performance

I8. IN BASKET EXERCISE, UK
W. C. F. Hartley, Management Games Ltd;
  Management Games Ltd
6th form and upwards
1-20; 2-3 hours
A non-interacting exercise on general
  management with a financial emphasis;
  decisions on personnel problems,
  company policy, strategy and finance

I9. IN-BASKET ON COMPLETED
  STAFF WORK; USA
A. Zoll, Addison Wesley
Management
1
Personnel management insights

I10. IN-BASKET ON MANAGE-
  MENT DEVELOPMENT: USA
A. Zoll; Addison-Wesley
Management
1
Shows that management development is
  the duty of every manager

I11. INDIAOMACY; USA
W. Moheisel; Albion
General
8
Red Indian tribal warfare, Diplomacy variant

I12. INDREL; UK
Management Games Ltd; Management
  Games Ltd
Management
Up to 10 individuals or up to 30 in teams;
  2-6 hours
Insight into industrial relations problems

I13. INDUSSIM (TOTAL INDUSTRY
  SIMULATION); 1970; USA
K. D. Ramsing; College of Business
  Administration, University of Oregon
College, graduate school, management
6-36; 1-4 hours
Creates environment for analysis and
  training of total firm

I14. INDUSTRIAL LOCATION
  GAME; 1972; UK
−; Community Service Volunteers
Secondary school
16
Insight into problems of planning and siting a
  modern steelworks; effects on community,
  etc.

I15. INDUSTRIAL MARKETING
  MANAGEMENT SIMULATION;
  1969; USA
J. J. Rath; School of Business Administration,
  Wayne State University
College, graduate school, management
5-50; 1 hour
Practice in allocating limited funds to
  business functions

I16. AN INDUSTRIAL RELATIONS
  GAME; 1973; UK
B. Beckett; Working Together
Students, all levels of management and
  workers
8-16; 2-6 hours
Practice in maximisation of resources in a
  competitive industrial situation

I17. INDUSTRIAL SALES
  MANAGEMENT GAME;
  USA
J. R. Greene, R. L. Sisson; Didactic Systems
  Inc; J. Wiley
Management trainees
1
Practice dealing with sales management
  problems

I18. INFLU; USA
−; Centre for Environmental Quality
  Management
Adults
Practice in developing immunisation
  programmes against influenza/pneumonia
  epidemics

I19. INFORCE; USA
−; Instructional Simulations Inc
Grade 9-12
Insight into law enforcement procedures

I20. THE INFORMATION GAME;
1969; USA
—; Training & Development Center
Management
2-16; 1-2 hours
Insight into failures in face-to-face
communications

I21. THE INLOGOV LOCAL
AUTHORITY GAME; 1968; UK
R. H. R. Armstrong; Birmingham
University
Local government
Insight into planning, budgeting, etc.

I22. INNER CITY PLANNING;
1970; USA
S. J. Fischer, R. A. Montgomery, J. J. Ziff;
MacMillan Co
High school, college
12-40; 3-5 hours
Insight into strategy and processes of inner
city living and urban renewal

I23. INNER CITY SIMULATION
LABORATORY: ABILITY
AND ACHIEVEMENT
TESTING TECHNIQUES;
1969; USA
D. R. Cruickshank; Science Research
Associates
Education students
2
Helping children to realise their capabilities
and limitations

I24. INNER CITY SIMULATION
LABORATORY: BO GREEN
DEFENDS HIS DIALECT;
1969; USA
D. R. Cruickshank; Science Research
Associates
Education students
2
Helping children who have limited
vocabulary and speech

I25. INNER CITY SIMULATION
LABORATORY: BARRY
PARSONS' AND MARK
CONNORS' REPORT CARDS;
1969; USA
D. R. Cruickshank; Science Research
Associates
Education students
2
Dealing with children who don't seem to
care about poor marks

I26. INNER CITY SIMULATION
LABORATORY: BO GREEN'S
MISTREATMENT AT HOME;
1969; USA
D. R. Cruickshank; Science Research
Associates
Education students
2
Dealing with children who are ill-treated at
home

I27. INNER CITY SIMULATION
LABORATORY: CLASSROOM
INTERRUPTIONS; 1969; USA
D. R. Cruickshank; Science Research
Associates
Education students
2
Dealing with classroom interruptions

I28. INNER CITY SIMULATION
LABORATORY: CONFERENCES
WITH MRS PARSONS
AND MRS CONNORS; 1969;
USA
D. R. Cruickshank; Science Research
Associates
Education students
2
Dealing with parents who are not interested
in their children's work

I29. INNER CITY SIMULATION LABORATORY: CRAIG POWERS, RECESS ROUGH-NECK; 1969; USA
D. R. Cruickshank; Science Research Associates
Education students
2
Dealing with a child who hurt others for no obvious reason

I30. INNER CITY SIMULATION LABORATORY: ELLEN ABRAMS' NERVOUSNESS; 1969; USA
D. R. Cruickshank; Science Research Associates
Education students
2
Helping nervous children

I31. INNER CITY SIMULATION LABORATORY: GETTING CLASS COOPERATION ON THE CENSUS; 1969; USA
D. R. Cruickshank; Science Research Associates
Education students
2
Helping inattentive children to remember and follow instructions

I32. INNER CITY SIMULATION LABORATORY: HAYWARD CLARK'S FEAR OF HIS FATHER; 1969; USA
D. R. Cruickshank; Science Research Associates
Education students
2
Helping a child upset by home conditions

I33. INNER CITY SIMULATION LABORATORY: MARSHA WRIGHT HAS AN EXCUSE; 1969; USA
D. R. Cruickshank; Science Research Associates
Education students
2
Dealing with a child who refuses to do, or gets out of, classwork

I34. INNER CITY SIMULATION LABORATORY: MARSHA WRIGHT'S MOTHER'S FRIEND: 1969; USA
D. R. Cruickshank; Science Research Associates
Education students
2
Dealing with children not motivated to work

I35. INNER CITY SIMULATION LABORATORY: MARY CHRISTIAN AND EMMA MARCH DISCUSS ASSIGN-MENTS; 1969; USA
D. R. Cruickshank; Science Research Associates
Education students
2
Helping children work independently

I36. INNER CITY SIMULATION LABORATORY: MORT COLEMAN, ACCEPT OR REJECT; 1969; USA
D. R. Cruickshank; Science Research Associates
Education students
2
Helping emotionally disturbed children

I37. INNER CITY SIMULATION LABORATORY: MRS WATERS' CLASS DOUBLES-UP; 1969; USA
D. R. Cruickshank; Science Research Associates
Education students
2
Handling excessively large classes

I38. INNER CITY SIMULATION LABORATORY: PANEL REPORT ON INDIVIDUAL-IZING INSTRUCTION; 1969; USA
D. R. Cruickshank; Science Research Associates
Education students
2
Finding time and methods to individualise instruction

I39. INNER CITY SIMULATION
LABORATORY: PHYLLIS
SMITH'S ABSENTEEISM;
1969; USA
D. R. Cruickshank; Science Research
Associates
Education students
2
Getting parents to cooperate on children's
attendance

I40. INNER CITY SIMULATION
LABORATORY: PHYLLIS
SMITH ASLEEP IN CLASS;
1969; USA
D. R. Cruickshank; Science Research
Associates
Education students
2
Dealing with children who come to school
without proper food or sleep

I41. INNER CITY SIMULATION
LABORATORY: PHYLLIS
SMITH'S HEARING PROBLEM;
1969; USA
D. R. Cruickshank; Science Research
Associates
Education students
2
Getting parents to take an interest in their
children's health

I42. INNER CITY SIMULATION
LABORATORY: PHYLLIS
SMITH SMELLS; 1969; USA
D. R. Cruickshank; Science Research
Associates
Education students
2
Interesting children in their personal
appearance and cleanliness

I43. INNER CITY SIMULATION
LABORATORY: Q-SORT
OF DISCIPLINE METHODS;
1969; USA
D. R. Cruickshank; Science Research
Associates
Education students
2
Finding satisfactory methods of disciplining
children

I44. INNER CITY SIMULATION
LABORATORY: RONALD
THURGOOD AND STANLEY
JONES RELATE PARENT
OPINIONS: 1969; USA
D. R. Cruickshank; Science Research
Associates
Education students
2
Dealing with parents who won't respond to
report cards or requests for conferences

I45. INNER CITY SIMULATION
LABORATORY: SHARON
STONE CALLS DEBBIE
WALKER A THIEF; 1969; USA
D. R. Cruickshank; Science Research
Associates
Education students
2
Helping children account for school supplies
and personal belongings

I46. INNER CITY SIMULATION
LABORATORY: SIDNEY
SAMS, DAYDREAMER; 1969;
USA
D. R. Cruickshank; Science Research
Associates
Education students
2
Helping daydreamers

I47. INNER CITY SIMULATION LABORATORY: SIDNEY SAMS LEAVES THE ROOM; 1969; USA
D. R. Cruickshank; Science Research Associates
Education students
2
Dealing with children misbehaving when left unsupervised for short periods

I48. INNER CITY SIMULATION LABORATORY: SIDNEY SAMS STRIKES OUT; 1969; USA
D. R. Cruickshank; Science Research Associates
Education students
2
Helping children with social adjustment problems

I49. INNER CITY SIMULATION LABORATORY: STANLEY JONES TESTS THE RULES; 1969; USA
D. R. Cruickshank; Science Research Associates
Education students
2
Dealing with children who want attention and will do anything to get it

I50. INNER CITY SIMULATION LABORATORY: TEACHING CHILDREN NOT PREPARED FOR GRADE-LEVEL WORK; 1969; USA
D. R. Cruickshank; Science Research Associates
Education students
2
Teaching children unprepared for grade-level work

I51. INNER CITY SIMULATION LABORATORY: THE GIRLS CLUB; 1969; USA
D. R. Cruickshank; Science Research Associates
Education students
2
Dealing with children who associate with children who are a bad influence

I52. INNER CITY SIMULATION LABORATORY: WESLEY BRIGGS ARRIVES EARLY; 1969; USA
D. R. Cruickshank; Science Research Associates
Education students
2
Handling children who will not obey the teacher

I53. INNER CITY SIMULATION LABORATORY: WESLEY BRIGGS BREAKS BRADLEY LIVESAY'S WATCH; 1969; USA
D. R. Cruickshank; Science Research Associates
Education students
2
Dealing with children who destroy other's property

I54. INNER CITY SIMULATION LABORATORY: WESLEY BRIGG'S AND THE CLASS'S LIBRARY BEHAVIOR; 1969; USA
D. R. Cruickshank; Science Research Associates
Education students
2
Dealing with children who rebel and try to upset the teacher

I55. INNER CITY SIMULATION
LABORATORY: WESLEY
BRIGGS MATCHES RONALD
THURGOOD; 1969; USA
D. R. Cruickshank; Science Research
Associates
Education students
2
Dealing with children who feel that stealing,
gambling, etc. are acceptable

I56. INNER CITY SIMULATION
LABORATORY: WESLEY
BRIGGS' MOM ASKS FOR
HELP; 1969; USA
D. R. Cruickshank; Science Research
Associates
Education students
2
Helping parents who say they cannot
control their children

I57. IN OTHER PEOPLES SHOES;
1972; UK
School Council; Longman
Secondary school, college
2; ½ hour
Insight into conflict in interpersonal
relations

I58. IN-QUEST; 1970; USA
R. W. Allen, L. E. Klapfer, R. L. Liss;
Didactic Systems Inc; International
Learning Corp
Junior high, high school
2-4; 2-8 hours
Practice in recognition of errors in
scientific enquiry

I59. INS (INTER-NATION
SIMULATION); USA
P. M. Burgess; Oregon State University
Students
Basic concepts of political decision-making

I60. INSIDE THE CITY; USA
—; High School Geography Project
High school
Insight into workings of inner city

I61. INSIGHT; 1967; USA
Games Research Inc; Instructional
Simulations Inc; Games Research Inc
General
2-20; 1½-3 hours
Tests powers of perception of personality

I62. INSITE; USA
—; School of Education, Indiana University
Student teachers
Practice in decision-making in the class-
room

I62. INSITE II; 1970; USA
K. L. Cyros, J. A. Langell; Massachusetts
Institute of Technology
University
Development of a computer-based space
information system

I64. INTEGRATED SIMULATION;
1968; USA
W. N. Smith, E. E. Estey, E. F. Vines;
South-Western Publishing Co
College, graduate school, management
4-96; 4-17 hours
Practice in integrating management
decisions in competition

I65. INTER-CITY TRANSPORT
POLICY GAME; 1972; UK
—; Imperial College
Transport policy and regional science
students
15
Insight into the aims, methods and
problems of public policy in the inter-
city transport field

I66. INTERCODE; UK
—; Griffin & George Ltd
Secondary school
Insight into processes of computers

167. THE INTER-COMMUNITY SIMULATION; 1968; USA
R. F. Goodman, E. Bonacich, J. F. Kirlin, A. Saltzstein; Department of Political Science, University of Southern California
College, graduate school
6; 6-12 hours
Demonstrates bargaining implications of public policy making

168. INTERNATIONAL OPERATIONS SIMULATION; 1964; USA
H. B. Thorelli; Free Press of Glencoe
Management
Total management game, operations in Brazil, USA and EEC

169. INTERLOCK; 1966; USA
A. A. Zoll; Addison–Wesley
College, management
Management of an aircraft instrument firm

170. INTERMEDIA; 1971; USA
B. D. Ruben, A. Talbot, L. Brown, H. La Brie; University Associates Press
Journalism and communications students
Practice in problem definition and solving and in interpersonal competance

171. INTER-NATION SIMULATION; 1966; USA
P. M. Burgess, L. E. Peterson, D. K. Benson; Ohio State University
College
Teaches concepts of political decision-making and international interaction in political, economic and military fields

172. INTER-NATION SIMULATION KIT; 1963; USA
H. Guetzkow, C. H. Cherryholmes; Science Research Associates
High school
24-50; 1½ hours
Insight into foreign relations and global politics

173. INTER-NATION SIMULATION KIT; 1963; USA
H. Guetzkow, C. H. Cherryholmes; Science Research Associates
College
A more advanced version of the high school kit

174. INTERNATIONAL; UK
P. Adolph; Subbuteo Ltd
General
Football

175. INTERNATIONAL PROCESSES SIMULATION; 1968; UK
P. Smoker; Lancaster University
University
Political simulation

176. INTERNATIONAL RELATIONS SIMULATION; USA
–; Michigan State University
College

177. INTERNATIONAL TRADE; 1968; USA
E. Rausch; Didactic Systems Inc; Science Research Associates
Grade 12 to college
6+; 2-5 hours

178. INTERVENTION; 1970; USA
–; Science Research Associates; Abt Associates
Grade 8 to junior college
Insight into decisions influencing expansion and involvement of the USA in small unstable nations after the Spanish-American War

179. INTERVIEWING; USA
E. Rausch; Didactic Systems Inc; Science Research Associates
Personnel management students
4+
Practice in selection interviewing

I80   INTOP (INTERNATIONAL
       OPERATIONS SIMULATION);
       1963
H. B. Thorelli, R. L. Graves, L. T. Howells;
Free Press of Glencoe
College, management
12-175
Problems of planning and co-ordinating in
widely spread decentralised operations;
demonstrates that different operating
environments require different operating
strategies

I81.   INTRODUCTION TO THE
        ENVIRONMENT; USA
–; Edcom Systems Inc
4th grade
Illustrates the relationship of human
societies to their geographical settings

I82.   INVASION; 1970; USA
–; Simulations Publications
General
Warfare in 1880's

I83.   INTRODUCTORY PANZER-
        BLITZ; USA
–; Avalon-Hill
General
World War II

I84.   IRON AND STEEL GAME; UK
B. P. Fitzgerald; St Mary's and St Paul's
Geographical Society
6-8; ¾-1 hour
Insight into industrial location at
three points in history

I85.   I-SPY; UK
E. Root; Rupert Hart-Davis
Pre-readers
Teaches recognition of initial vowels and
consonants

I86.   ITALY; USA
–; Simulations Publications Inc
General
Italian campaign in World War II

J1.   THE JANUS JUNIOR HIGH
       PRINCIPALSHIP; USA
–; University Council for Educational
Administration
Educational administrators

J2.   JEFFERSON TOWNSHIP; USA
–; University Council for Educational
Administration
Educational administrators

J3.   JESAT; UK
–; Satla Products
General
Teaches spatial relationships

J4.   JIGABRIX; 1971; UK
A. Ighodaro, M. Ford; Think Games Ltd
4-7 years old

J5.   JIG-SAW CARDS; 1937; UK
F. J. Schonell; Oliver & Boyd Ltd
Primary school
Practice in number bonds

J6.   JOBLOT (A PRODUCTION
       MANAGEMENT GAME);
       1970; USA
G. Churchill; MacMillan Co
College, graduate school, management
2+; 3-5 hours
Practice in group work to deal with
industrial environment; insight into
decision-techniques

J7.   JOE BAILEY ACTION MAZE;
       1966; USA
A. A. Zoll; Addison Wesley Ltd
Management
Insight into a personnel supervisory problem

J8   THE JOHN AND GEORGE
      INTERVIEW: (PART OF
      SUPERVISORY SKILLS
      SERIES); 1969; USA
–; Training Development Center
Management
2-16; 1-2 hours
Demonstrates interviewing techniques:
probing for information, discriminating
between a response and a reply

J9.   JOHN STEVENS; UK
−; Young Mens Christain Association
Identifies attitudes, makes group face
 problems

J10.   JOURNEYS OF ST PAUL;
  USA
−; Avalon-Hill Inc
General
St Paul's journeys

J11.   JULIUS CAESAR; 1969; USA
R. W. McElhave; Pleasant Hill High School
High school
10-100; 2-4 weeks
Presents the play 'Julius Caesar'

J12.   JUMBO ANIMAL DOMINOES;
  1972; UK
−; Habitat Ltd
3-8 years old

J13.   JUMBOLINO; Holland
−; Jumbo Games
Primary school
6+
Teaches number bonds

J14.   JUNIOR ELECTRO; Holland
−; Jumbo Games
Slow readers
Increases reading ability

J15.   JUMPIN; USA
−; 3M Company
General
2 (or 2 teams); 20-40 minutes
Tactical board game

J16.   JUTLAND; 1967; USA
−; Avalon-Hill Inc
General
2+; 3-6 hours

Naval tactics in World War 1;
 practice in strategic decision-
 making

J17.   JUVENILE DELINQUENCY
  GAME; USA
H. C. Olafson, D. Callaghan; Research
 Analysis Corp
Children, parents, teachers, law officers
Insight into values, attitude change and
 behavioural change

K1.   KALAHA; Denmark
−; Piet Hein
General

K2.   KIDDIE KARDS; USA
N. Cocanower; Antioch Bookplate Co
4-10 years old
2-6; 6-25 minutes
Teaches word recognition and word/picture
 matching

K3.   KI PRACTICE; 1962; USA
L. E. Allen; Wff'n Proof Inc
Grade 1 to college
2+
Teaches use of the conjunction in rule of
 logic

K4.   KIDDY SNAPS; UK
−; Spears Games
Children
Visual discrimination practice

K5.   KIKUYU DIORAMA; USA
−; Ed/Com Systems Inc; Abt Associates Inc
High school
Life in Kenya

K6.   THE KING VERSUS THE
  COMMONS; USA
−; KDJ Instructional Systems Inc.;
  Educational Development Center
Grade 8
Insight into the confrontation between
 Charles I and Cromwell

K7.   KINGS GAME; 1644; Holland
C. Weikhmann; −
Military
2
Practice in strategic and tactical thinking

K8. KIO PROOF; 1962; USA
L. E. Allen; Wff'n Proof Inc
Grade 1 to college
2+
Practice in deduction using Ki and Ko rules

K9. KIO-CIO-R. PROOF; 1962; USA
L. E. Allen; Wff'n Proof Inc
Grade 1 to college
2+
Practice in deduction and construction of
logical systems

K10. KIO-CIO-R-AIO PROOF;
1962; USA
L. E. Allen; Wff'n Proof Inc
Grade 1 to college
2+
Practice in deduction and construction of
logical systems

K11. KIO-CIO-R-AIO-NIO PROOF;
1962; USA
L. E. Allen; Wff'n Proof Inc
Grade 1 to college
2+
Practice in deduction and construction of
logical systems

K12. KIO-CIO-R-AIO-NIO-EIO
PROOF; 1962; USA
L. E. Allen; Wff'n Proof Inc
Grade 1 to college
2+
Practice in deduction and construction of
logical systems

K13. KIO-CO PROOF; 1962; USA
L. E. Allen; Wff'n Proof Inc
Grade 1 to college
2+
Practice in deduction using KI, KO and CO
rules and construction of logical systems

K14. KNOW YOUR STATES;
1956; USA
E. W. Dolch; Garrard Publishing Co
Grade 4 and upwards
2-4; 10-15 minutes
Teaches recognition, pronunciation and
spelling of US states' names and rules
for syllable division of words

K15. KOLKHOZ; USA
−; Abt Associates Inc
Junior high school
Insight into economic philosophy of the
collective farm

K16. KOLOR KRAZE; UK
−; Waddingtons Ltd
Primary school and upwards
1
Teaches topology

K17. KONFUSION; UK
C. Hemming; C. Hemming
General
Diplomacy variant

K18. KO PRACTICE; 1962; USA
L. E. Allen; Wff'n Proof Inc
Grade 1 to college
2+
Teaches use of the conjunction out rule of
logic

K19. KOREA: CAMPAIGN GAME;
1970; UK
−; Simulations Publications
General

K20. KOREA; INTERVENTION;
1970; UK
−; Simulations Publications
General

K21. KOREA: INVASION; 1970; UK
−; Simulations Publications
General

K22. KOREA: STALEMATE;
1970; UK
−; Simulations Publications
General

K23. KRIEGSPIEL; USA
—; Avalon-Hill Inc
General
Practice of strategies, tactics, logistics and
operations of war and diplomacy in
abstract form

K24. KURSK; USA
—; Simulations Publications Inc
General
Battle of Kursk 1943

L1. THE L GAME; 1967; UK
E. de Bono; Penguin Books
Children and adults
2
Practice of spatial skills

L2. LABORATORY EXERISE ON
SELECTION; USA
K. F. McIntyre; University Council for
Educational Administration
Educational administrators

L3. LABOR VS MANAGEMENT;
1966; USA
J. D. Gearon; Social Education
High school, college
Insight into labour/management relations and
clarification of labour history

L4. LADDER GAME; UK
—; Educational Supply Association
Primary school
2
Practice in counting

L5. LAND AND WATER ANIMALS
GAME; 1967; USA
H. A. Springle; Science Research Associates
Pre-school to grade 1
1-4; 20-30 minutes
Insight into 'more than' and 'less than'

L6. LANDING FORCE WAR
GAME; 1958; USA
—; US Marine Corps
Military
2; 4 months
Insight into Cuban situation tactics

L7. THE LAND USE TRANS-
PORTATION SIMULATION;
1967; UK
D. Macunovitch; Planning & Transport
Research & Computation Ltd
Postgraduate, town planners
13-36; 3-4 days
Insight into cause and effect relationships
involved in community decision-making
chains

L8. LANGOFUN; UK
—; Langofun International
General
Travel from London to Monte Carlo;
French language practice

L9. LANGUAGE LOTTO-ACTIONS;
USA
L. G. Gotkin; New Century
Pre-school, grade 1
Teaches one to describe actions in complete
sentences

L10. LANGUAGE LOTTO—
COMPOUND SENTENCES;
USA
L. G. Gotkin; New Century
Pre-school, grade 1
Teaches use of conjunctions

L11. LANGUAGE LOTTO—
MORE ACTIONS; USA
L. G. Gotkin; New Century
Pre-school, grade 1
Practice in use of verbs in sentences

L12. LANGUAGE LOTTO—
OBJECTS; USA
L. G. Gotkin; New Century
Pre-school, grade 1
Teaches vocabulary

L13. LANGUAGE LOTTO—
PREPOSITIONS; USA
L. A. Gotkin; New Century
Pre-school, grade 1
Teaches use of prepositions

L14. LANGUAGE LOTTO–
RELATIONSHIPS; USA
L. G. Gotkin; New Century
Pre-school, grade 1
Practice in use of relationships to explain
processing

L15. LAW ENFORCEMENT AND
CRIMINAL JUSTICE
PROBLEMS; USA
R. M. Longmire, R. W. Rae; Research
Analysis Corporation
Management
3-40; ¾-1 hour
Focuses planning problems in justification
of requirements, completeness and
validity within safety requirements

L16. LAWYER; UK
–; Inter-Action
General
Insight into problems of self-control and
liberation

L17. LEADERSHIP EXERCISE; UK
–; Glamorgan Youth Service
Youth leaders
Practice in decision-making in a youth club

L18. LEADERSHIP GAME; 1968; USA
R. E. Ohm; University Council for
Educational Administration
Graduate school
Insight into advantages of rationally-based
decision-making

L19. THE LEADERSHIP GAME;
1970; UK
F. Treuhertz; Association for Jewish Youth
Students
Practice in decision-making in a Jewish
Youth Club

L20. LEADERSHIP OF NONSUCH
YOUTH CLUB; UK
A. Aldrich; Wiltshire Training Agency
Youth workers
Trains youth leaders

L21. LEADING GROUPS TO
BETTER DECISIONS; USA
E. Rausch; Didactic Systems Inc
Management
3-5
Practice in problem-solving and decision-
making for effective conference leadership

L22. LEAGUE CHAMPIONSHIP; UK
–; Sports-Link Ltd
General
Up to 6
Association football

L23. A LEARNING DISABILITY
SIMULATION–1971; USA
F. W. Broadbent, R. Meehan; Syracuse
University
In-service teachers
1; 6 hours
Teaches skills for dealing with children
with mild to moderate learning problems

L24. THE LEBLING VARIANT;
USA
D. Lebling; D. Lebling
General
Diplomacy variant

L25. LEIPZIG; USA
–; Simulations Publications
General
Napoleon's 1813 European campaign

L26. LE MANS; USA
–; Avalon-Hill Inc
10 years old and over
2-12
Insight into rally driving strategy

L27. LETTO; USA
–; Houghton Mifflin
Pre-reading
2+
Practice in letter recognition

L28. LEXICON; UK
–; Waddingtons Ltd
General
2-6; 30 minutes
Practice in spelling and vocabulary exercise

L29.   THE LIB GAME; 1971; USA
—; Creative Communications and Research
College, adults
4+; 1½-3 hours
Insight into dynamics of Women's Lib
    movement and reactions to it

L30.   LIBERTE; USA
—; Interact
Grade 7 to college
1-35
Insight into French Revolution and the
    processes leaving up to it

L31.   LIE, CHEAT AND STEAL; USA
—; Dynamic Design Industries
General
Competition for political office

L32.   LIFE CAREER; USA
S. S. Boocock; Western Publishing Co;
    Academic Games Associates
Junior high, high school
2-20; 1-6 hours
Insight into labour, education and marriage
    markets of the USA

L33.   LIFE IN ISRAEL; 1970; USA
Creative Studies Inc; Creative Studies Inc;
    The Jewish Agency
College, graduate school, emigrants to
    Israel
3-20; 2-5 hours
Insight into the problems of absorption into
    Israel in terms of employment, housing and
    education

L34.   LIFELINE: IN OTHER
        PEOPLES SHOES; 1972; UK
—; Longman
Secondary school
Teaches moral standards

L35.   LIFELINE: PROVING THE
        RULE; 1972; UK
—; Longmans
Secondary school
Teaches moral standards

L36.   LIFELINE: WHAT WOULD
        YOU HAVE DONE; 1972; UK
—; Longmans
Secondary school
Teaches moral standards

L37.   LISTEN; USA
P. Lamb; Lyons and Carnahan
Grades 1-6
3-4
Teaches recognition and spelling of common
    beginning consonant digraphs and
    clusters

L38.   LISTEN, MARK AND SAY;
        1967; USA
L. G. Gotkin, S. Mason, E. Richardson;
    Appleton-Century-Crafts
Infant and primary school
1-30; 15 minutes +
Teaches reading readiness, skills, listening
    comprehension, language facility,
    reasoning, problem-solving, test-taking

L39.   THE LISTENING GAME
        (THE MARSHALL-DUNNE
        CONTROVERSY); 1969; USA
—; Training Development Center
Management
6-16; 1-2 hours
Practice in listening in depth, perceiving
    what is said and what is meant, examining
    the influence of mental set

L40.   LITTLE WARS; 1912; UK
H. G. Wells; —
Children

L41.   LOCAL AUTHORITY GAME;
        1970; UK
R. H. R. Armstrong; University of Birmingham
College, local government
Insight into major factors in town develop-
    ment and associated problems facing
    local authority

L42. LOCATING A COSMETICS FACTORY IN THE UK; 1971; UK
J. P. Cole; Geography Department, University of Nottingham
College, management
4; ½ hour
Insight into the processes of factory location

L43. LOCATING VEHICLE SALES AND SERVICE CENTRES IN AMNESIA; 1971; UK
J. P. Cole; Department of Geography, University of Nottingham
College, management
Industrial location

L44. LOCAL GOVERNMENT REFORM; 1971; UK
J. P. Cole; Geography Department, Nottingham University
Secondary school upwards
Finding the lowest cost new system of local government units and their capitals in two countries

L45. THE LONDON GAME; UK
—; Condor
Primary, secondary school
2-5
Teaches geography of London and places of interest via the underground

L46. LONELINESS; UK
P. J. Tansey; Berkeshire College of Education
Education students
2+
Insight into problems and backgrounds of lonely people in a college of education

L47. LOST BATTLES; USA
—; Simulations Publications Inc
General
Division level tactical game

L48. LOW BIDDER: THE GAME OF MANAGEMENT STRATEGY; 1964; USA
W. R. Park; Entelek Inc; Didactic Systems Inc
High school, college, contractors, builders
2-25; 1 hour
Teaches strategy of selection for a job with few bidders; estimating mark-up, protection against cut-throat competition

L49. LUCKY DUCK; USA
P. Lamb; Lyons and Carnahan
Grades 1-6
3-6
Teaches short vowel sounds, phonemic discrimination

L50. LUDO; 1972; UK
B. Sampson; Galt Toys
4½-7 years old
Teaches number

L51. LUFTWAFFE; USA
—; Avalon-Hill Inc
General
1
Practice in planning and adapting strategies in World War II air war context

L52. LUGS (LAND USE GAMING SIMULATION); 1968; UK
J. L. Taylor, R. N. Maddison; Urban Affairs Quarterly
University
Land use

L53. THE LUMBER YARD GAME; 1969; USA
—; Training Development Center
Management
15-20; 1-2 hours
Teaches principles of training; entry and terminal behaviours; the role of classroom v. on-the-job training

M1.   MACHIAVELLI; 1972; UK
J. P. Cole; Department of Geography,
   Nottingham University
Secondary school upwards
5
Struggle for territory requiring skill in
   organisation of armies, formation of
   alliances and appreciation of odds in
   attacking

M2.   MACHINIST; USA
E. Gollay; Harvard University; US Office
   of Education; Abt Associates
High school
Insight into career selection

M3.   MARROTOPIA; UK
J. P. Cole; University of Nottingham
Secondary school
5-7; 1-2 hours
World airlines game showing complexity of
   negotiations and transactions

M4.   MADISON ASSISTANT
   SUPERINTENDANT FOR
   BUSINESS MANAGEMENT;
   −1967; USA
−; University Council for Educational
   Administration
Management
Up to 25
Practice in application of administrative
   theories to analysis and solution of
   problems

M5.   MADISON ASSISTANT
   SUPERINTENDENT FOR
   INSTRUCTIONAL SERVICE;
   −1967; USA
−; University Council for Educational
   Administration
Management
1-25
Application of administrative theories to
   analysis and solution of problems

M6.   MADISON PUBLIC SCHOOLS:
   ADMINISTRATION OF
   VOCATIONAL EDUCATION;
   USA
−; University Council for Educational
   Administration
Educational administrators

M7.   THE MADISON PUBLIC
   SCHOOL SUPERINTENDENCY;
   1966; USA
G. Immegart, P. Rusche, A. Benyenuto;
   University Council for Educational
   Administration
Educational administrators
Practice in experience of the role of
   assistant school superintendant

M8.   MADISON SECONDARY
   PRINCIPALSHIP; 1966; USA
D. Anderson, H. Laughlin; University
   Council for Educational Administration
Educational administrators, college
1-25
Practice in problem-solving of the kind done
   by secondary school heads

M9.   MAGIC NUMBER; 1967; USA
M. B. Lorenzana; Motor Products Corp
9 years old and over
110 minutes
Develops accuracy in arithmetic addition

M10.   MAH-JONGG; UK
−; J. Jacques & Son Ltd
General

M11.   MAKE-A-FACE; UK
−; Spears Games
All ages
Teaches spatial and colour recognition

M12.   MAKE-IT; UK
E. Root; Rupert Hart-Davis
Pre-readers
Teaches word making

M13.   MAKE ONE (A FRACTION
       GAME); 1956; USA
E. W. Dolch; Garrard Publishing Co
Grade 3
2-5; 10-15 minutes
Teaches fractions and percentages, their
   equivalence and addition

M14.   MAN: A STUDY IN ADAP-
       TATION; USA
−; Ed/Com Systems Inc
Secondary school

M15.   MAN AND TOWNS; UK
R. Pepper, A. Calland; Jackdaw Publications
Secondary school
includes a town planning game

M16.   MAN IN HIS ENVIRON-
       MENT; 1971; UK
−; Coca-Cola Export Corporation
Schools
10-70; 1 hour +
Insight into the inter-related effects of
   actions on our environment

M17.   MAN POWER; 1969; USA
M. Hanan; Hanan & Son
Management
1+; 30-60 minutes
Practice in sales marketing management,
   organisation performance

M18.   MANAGEMENT; USA
−; Didactic Systems Inc
College, management
2-4
Practice in analysing market conditions
   and competitors circumstances, operation
   of factories and warehouses, seeking
   efficiency

M19.   MANAGEMENT; 1958; USA
−; Avalon-Hill Inc
High school, college
2-4; ½-2 hours
Insight into process of management

M20.   MANAGEMENT BY
       OBJECTIVES; 1969; USA
−; Didactic Systems Inc
College, management
3+; 2-2½ hours
Practice in alternative strategies leading
   to MBO

M21.   MANAGEMENT BY
       OBJECTIVES; 1970; UK
R. H. R. Armstrong, M. Hobson; Institute of
   Local Government Studies, Birmingham
   University
Post-experience, post-graduate, local
   government
4+; 2 hours +
An application of NEXUS (q.v.) to MBO

M22.   MANAGEMENT DECISION-
       MAKING LABORATORY;
       1959; USA
G. T. Hunter; IBM
College, management
12-15; 1 day
Demonstration of interaction of management
   areas; illustration of fundamental
   relationships and terminology of
   management and concept of simulation

M23.   MANAGEMENT DECISION
       SIMULATION; 1960; USA
S. Vance; McGraw-Hill Inc
College, graduate school, management
15-32; 5 hours
Insight into complexity of industry, cause
   and effect nature of decision-making and
   performance and the meaning of counter
   forces in corporate decision-making

M24.   MANAGEMENT DECISION
       SIMULATION (COMPUTER
       VERSION); 1960; USA
S. C. Vance, C. F. Gray; School of Business
   Administration, University of Oregon
College, graduate school, management
3+; 5 hours
As in 'Management Decision Simulation' (q.v.)

M25. MANAGEMENT FOR
SUPERVISORS; 1970; USA
–; Didactic Systems Inc
College, supervisors and foreman
3+; 2-3½ hours
Practice in management skills

M26. THE MANAGEMENT GAME;
1970; USA
F. W. McFarlan, J. L. McKenney, J. Soilec;
Macmillan Co
Graduate school, management
20; 3-10 days
Development of functional integration,
integration of time-sharing models,
encouragement of strategy formation,
human relations experience

M27. MANAGEMENT OF THE
PHYSICAL DISTRIBUTION
FUNCTION; 1969; USA
–; Didactic Systems Inc
College, management
3+; 2-3½ hours
Practice in decisions on warehouse location;
inventory distribution, function
responsibilities, motivation of
employees

M28. MANAGEMENT SIMULATION
AND SEMINAR; USA
R. N. Stapleton; Western Center
High school, college, graduate school,
military, management
12-100; 40 hours
Encourages personal concern with the
human side of enterprise; self-insight
through feedback from team

M29. MANAGEMENT: THE
MANUFACTURING AND
INDUSTRIAL ENGINEERING
FUNCTIONS
–; Didactic Systems Inc
Management trainees
3+
Framework for practice in decision-making

M30. MANAGING A CHURCH
HALL AND PARISH YOUTH
WORK; UK
A. Aldrich; Wiltshire Training Agency
Youth leaders, theological students,
ministers
Practice in decisive leadership, co-operation;
insight into need for organisation

M31. MANAGING THE ENGINEER-
ING FUNCTION; USA
–; Didactic Systems Inc
Management
3-5
Practice in decision-making in departmental
organisations, motivating engineering
workers, setting goals

M32. MANAGING THE QUALITY
CONTROL FUNCTION; USA
–; Didactic Systems Inc.
Management
3+
Practice in decision-making

M33. MANAGING THE WORKER;
1970; USA
W. Archey, J. J. Ziff, A. Walker;
McMillan Co
College, management
10-20; 3-6 hours
Insight into necessity for sensitivity to
workers' needs, methods of providing
atmosphere for subordinates development and self-actualization

M34. MANCHESTER; USA
M. Rosen, J. Blaxall; Abt Associates Inc,
Education Development Center
High school
Illustrates economics of early Industrial
Revolution in Britain, showing migration
to towns

M35. MANEX; UK
R. Boardman; University of Surrey
University
Insight into international crisis

M36. MANOVRA SULLA CARTA;
1878; Italy
Naval
Naval tactics and strategy

M37. MANSYM; 1965; USA
R. E. Schellenberger; Management
   Development Inc
College, graduate school, management
4-5; 2 months
Practice in dealing with management
   problems, roles and systematic techniques

M38. A MANUAL FOR
   CONDUCTING PLANNING
   EXERCISES; 1969 USA
P. A. Twelker; Teaching Research
High school
16-36; 150-250 minutes
Teaches a general format for preparing and
   conducting a planning exercise

M39. MANUFACTURING AND
   INDUSTRIAL ENGINEERING
   MANAGEMENT; 1970; USA
—; Didactic Systems Inc
College, management
3+; 2-3½ hours
Practice in decision-making in priority
   setting, concept estimates, operations
   improvement, phasing-in new designs,
   deciding standards, applying learning
   curves

M40. MANUFACTURING
   MANAGEMENT SIMULATION;
   USA
—; American Management Association
Management

M41. MANUFACTURING
   MANAGEMENT SIMULATION;
   USA
—; Sperry Rand Corp
Management

M42. MANUFACTURING
   MANAGEMENT SIMULATOR;
   1968; USA
D. D. McNair, A. P. West; Simulated
   Environments Inc
High school, management
8-60; 1½-2 hours
Practice decision-making skills and learn
   advertising management theory

M43. MAP MAKER; 1970; Canada
D. R. Olson; D. R. Olson
Primary school
1; 10 minutes +
Development of spatial ability and
   perceptual-motor skills

M44. MARIRAMA; 1973; USA
—; Marirama Inc
General
Board game based on marijuana smuggling

M45. THE MARKET; 1968; USA
E. Rausch, H. J. Cranmer; Science
   Research Associates; Didactic Systems Inc
High school, college
3+; 2-5 hours
Demonstrates how price is established through
   changing demand and supply conditions,
   concept of diminishing returns

M46. MARKET; USA
W. Rader, K. Chapman, L. Orcear; Abt
   Associates Inc; University of Chicago
Grade 6
18-40; 40-50 minutes
Insight into price determination in a market
   economy

M47. MARKET GAME; USA
—; Holt, Rhinehart and Winston
Grade 6
4
Insight into free market process

M48. MARKET GAME; 1967; USA
—; Joint Council for Economic Education
Junior high school
20-50; 30 time periods
Insight into price establishment in a market
   economy

M49.　MARKET NEGOTIATION
　　　MANAGEMENT GAME; USA
J. R. Greene, R. L. Sisson; Didactic
　Systems Inc; J. Wiley & Son
Management
5+
Insight into interdependence of firms in a
　distribution chain

M50.　MARKET PLACE; USA
−; Edcom Systems Inc; Abt Associates Inc
Grade 4
Illustrates the economics of an African
　market-place via bartering of food, hard-
　ware, livestock and jewelry

M51.　MARKET PLAN; 1965; USA
M. Hanan; Hanan & Son
Management
15-45; 7 hours
Training in management knowledge, skill
　in marketing management

M52.　MARKET PLANNING; USA
−; Didactic Systems Inc
Management students
3+
Practice in preparation of marketing plans

M53.　MARKET TEST; 1969; USA
M. Hanan; Hanan & Son
Management
1+; 30-60 minutes
Practice of new product strategy and test-
　marketing performance

M54.　MARKETING MANAGEMENT
　　　EXERCISE; 1965; UK
J. W. Fairhead, D. S. Pugh, W. J. Williams;
　E.U.P.
Management
Insight into problems of using techniques of
　marketing and market research; practice
　in displaying marketing resources

M55.　MARKETING A NEW
　　　PRODUCT; 1970; USA
J. Zif, I. Ayal, E. Orbach; McMillan Co
College, management
4-24; 6 hours +
Practice of skills in analytical use of market
　research data, strategic and tactical
　decision-making; reinforces analytical
　thinking

M56.　MARKSIM; A MARKETING
　　　DECISION SIMULATION; USA
P. S. Greenlaw, F. W. Kniffin; Didactic
　Systems Inc; International Textbook Co
College, management
2-150; 2 hours
Practice in marketing decision-making

M57.　MARNE 1914; USA
−; Simulations Publications Inc
General
First World War

M58.　MASTER MIND; UK
−; Invicta Plastics
General
General knowledge

M59.　MASTERPIECE; UK
−; Palitoy Ltd
8 years old and over
3-7; ¾ hour
Gives insight into auctioneering

M60.　MATCH; 1953; USA
E. W. Dolch; Garrard Publishing Co
Kindergarten, grade 1
1-2; 10-15 minutes
Practice of ability to distinguish similarities
　and differences between words, and
　immediate recognition of the most
　common nouns

M61.　MATCH; USA
P. Lamb; Lyons and Carnahan
Grades 1-6
3-6
Teaches discrimination of long vowel and
　dipthong sounds; associates them with
　spelling patterns

M62. MATCH-IT; UK
E. Root; Rupert Hart-Davis
Pre-readers
Teaches sound combinations

**M63. MATERIAL MANAGEMENT
SIMULATION; 1966; USA**
A. A. Zoll; Addison-Wesley
Management
Purchasing management practice

**M64. MATERIALS INVENTORY
MANAGEMENT GAME; USA**
J. R. Greene, R. L. Sisson; Didactic
Systems Inc; J. Wiley & Sons
Management
3+
Demonstrates use of 'Economic Order
Quantity' formula and simple demand
forecasting

**M65. MATERIALS MANAGEMENT
SIMULATION; USA**
–; American Management Association
Management

**M66. MATERIALS MANAGEMENT
SIMULATION; USA**
–; University of Tennessee
Management

M67. MATH BLOCKS; 1971; USA
–; Wff'n Proof; Mills Center
Pre-school
Practice of number and four rules

M68. MATH MAGIC; 1954; USA
E. E. Mazer, D. R. Bolton; Cadaco Inc
5-10 years old
2-4
Practice in multiplication and factoring

M69. MATH MATRIX 1; 1971; USA
L. G. Gotkin; Appleton-Century-Crofts
Infants
4-7; 10-20 minutes
Teaches recognition of red, yellow, blue
and green, identification of trees,
buildings and cranes, discrimination
between short, medium, tall and shorter,
shortest, taller, tallest.

M70. MATH MATRIX 2; 1971; USA
L. G. Gotkin; Appleton-Century-Crofts
Infants
4-7; 10-20 minutes
Teaches identification of pictures of birds,
cows, elephants, cherries, oranges, water
melons, bicycles, cars and buses,
discrimination between small, medium
and large and smaller, smallest, larger,
largest

M71. MATH MATRIX 3; 1971; USA
L. G. Gotkin; Appleton-Century-Crofts
Infants
4-7; 10-20 minutes
Teaches identification of pictures of jungle
gyms, slides and fences, discrimination
between top, middle and bottom, on
top, in the middle and on the bottom
and above and below

M72. MATH MATRIX 4; 1971; USA
L. G. Gotkin; Appleton-Century-Crofts
Infants
4-7; 10-20 minutes
Teaches identification of pictures of plates
of food, glasses of milk and bags of
potato crisps, discrimination between
full, half-full and empty, most and least
and more than and less than

M73. MATH MATRIX 5; 1971; USA
L. G. Gotkin; Appleton-Century-Crofts
Infants
4-7; 10-20 minutes
Teaches how to identify pictures of streets,
classrooms and bedrooms, and how to
interchange not a single one, nobody
and no one, lots of people, many people,
some people and not many people

M74. MATH MATRIX 6; 1971; USA
L. G. Gotkin; Appleton-Century-Crofts
Infants
4-7; 10-20 minutes
Teaches identification of pictures of birds
in nest, squirrels on branch and bees
on flower; counting 0, 1, 2 and 3 and
identification of groups of 0, 1, 2 or 3
objects

M75. MATH MATRIX 7; 1971; USA
L. G. Gotkin; Appleton-Century-Crofts
Infants
4-7; 10-20 minutes
Teaches how to identify pictures of
children with rain-wear and swim-wear
and discrimination between enough,
none and too many

M76. MATH MATRIX 8; 1971; USA
L. G. Gotkin; Appleton-Century-Crofts
Infants
4-7; 10-20 minutes
Teaches how to identify pictures of rabbits,
pigs, deer and squirrels with food; carrots,
ears of corn, apples and acorns;
discrimination between more than, fewer
than, enough; substitution of right
amount/enough, not as many/fewer than,
to much/more than

M77. MATH MATRIX 9; 1971; USA
L. G. Gotkin; Appleton-Century-Crofts
Infants
4-7; 10-20 minutes
Teaches identification of pictures of firemen,
postmen, doctors, policemen and related
objects; counting to 2, 3, 4 and 5;
identification of groups of 2, 3, 4 and 5
objects and substitution of same number
as/as many as

M78. MATH MATRIX 10; 1971; USA
L. G. Gotkin; Appleton-Century-Crofts
Infants
4-7; 10-20 minutes
Teaches identification of pictures of toys,
writing instruments and tools; counting
4, 5, 6 and 7, identification of groups
of 4, 5, 6 and 7 objects; discrimination
between more than/less than

M79. MATH MATRIX 11; 1971; USA
L. G. Gotkin; Appleton-Century-Crofts
Infants
4-7; 10-20 minutes
Teaches identification of pictures of sweets,
containers, fruits and vegetables;
counting 7, 8, 9 and 10, their
identification as groups and counting
10 to 0

M80. MATH MATRIX 12; 1971; USA
L. G. Gotkin; Appleton-Century-Crofts
Infants
4-7; 10-20 minutes
Reviews all mathematics learned in games
1 to 9

M81. MATH MATRIX 13; 1971; USA
L. G. Gotkin; Appleton-Century-Crofts
Infants
4-7; 10-20 minutes
Teaches identification of children standing
in line at telescopes and water fountains
and discrimination between 1st, 2nd and
3rd and last, just before and just after

M82. MATH MATRIX 14; 1971; USA
L. G. Gotkin; Appleton-Century-Crofts
Infants
4-7; 10-20 minutes
Teaches identification of pictures of lions,
monkeys and penguins, eating working
and playing and of numerals 1, 2 and 3

M83. MATH MATRIX 15; 1971; USA
L. G. Gotkin; Appleton-Century-Crofts
Infants
4-7; 10-20 minutes
Teaches identification of pictures of houses
with diamond shape on the door and
contents in windows and discrimination
of numerals 2, 3, 4 and 5

M84. MATH MATRIX 16; 1971; USA
L. G. Gotkin; Appleton-Century-Crofts
Infants
4-7; 10-20 minutes
Teaches identification of circus scenes and
numerals 5, 6, 7 and 8

M85. MATH MATRIX 18; 1971; USA
L. G. Gotkin; Appleton-Century-Crofts
Infants
4-7; 10-20 minutes
Teaches identification of pictures of space
travel etc. and identification of numerals
8, 9 and 10

M86.   MATH MATRIX 18; 1971; USA
L. G. Gotkin; Appleton-Century-Crofts
Infants
4-7; 10-20 minutes
Teaches identification of pictures of people
with fishing rods, balloons and bats and
how to interchange, i.e. adding one/one
more

M87.   MATH MATRIX 19; 1971; USA
L. G. Gotkin; Appleton-Century-Crofts
Infants
4-7; 10-20 minutes
Teaches identification of pictures of kittens,
dogs, rabbits, and how to interchange,
i.e. take one away/one less

M88.   MATH MATRIX 20; 1971; USA
L. G. Gotkin; Appleton-Century-Crofts
Infants
4-7; 10-20 minutes
Teaches how to identify pictures of pirate
ships and related objects; reviews the
contents of all the previous games

M89.   MATH PACK; 1969; USA
–; Community Makers
Primary, junior high school
1-5; 10-60 minutes
Practice problem reading and arithmetic
skills to give language/number system
equivalences

M90.   MATRIX; 1962; USA
J. W. Plattner, L. W. Herron; Procter and
Gamble
Management
Teaches purchasing management

M91.   MATRIX GAME 1; 1967; USA
L. G. Gotkin; New Century
Infants
1 day
Gives practice in following simple directions
with shapes and of stating what is done

M92.   MATRIX GAME 2; 1967; USA
L. G. Gotkin; New Century
Infants
1 day
Gives practice in choosing a shape, a picture
and both together and in saying what was
done

M93.   MATRIX GAME 3; 1967; USA
L. G. Gotkin; New Century
Infants
1 day
Gives practice in describing pictures and
following two directions

M94.   MATRIX GAME 4; 1967; USA
L. G. Gotkin; New Century
Infants
1 day
Gives practice in giving directions

M95.   MATRIX GAME 5; 1967; USA
L. G. Gotkin; New Century
Infants
1 day
Reviews material in games 1-4

M96.   MATRIX GAME 6; 1967; USA
L. G. Gotkin; New Century
Infants
1 day
Gives practice in abstracting similar
characteristics

M97.   MATRIX GAME 7; 1967; USA
L. G. Gotkin; New Century
Infants
1 day
Gives practice in abstracting similar
characteristics

M98.   MATRIX GAME 8; 1967; USA
L. G. Gotkin; New Century
Infants
1 day
Teaches how to abstract complex
characteristics and classify them

M99. MATRIX GAME 9; 1967; USA
L. G. Gotkin; New Century
Infants
1 day
Teaches negation of shape and picture
labels

M100. MATRIX GAME 10; 1967;
USA
L. G. Gotkin; New Century
Infants
1 day
Gives practice of deducing missing
characteristics

M101. MATRIX GAME 11; 1967;
USA
L. G. Gotkin; New Century
Infants
1 day
Teaches the abstraction of common
characteristics

M102. A MAYOR FOR MOUNT
VAN BUREN; USA
J. D. Gearon; Chicago Public School
Systems
High school
Insight into local politics

M103. MAZE; UK
—; Mosesson Games Ltd
General
1+
Gives manipulative practice

M104. MEDPLAN; USA
—; Abt Associates
Health officials
Selection of goals and priorities, development
of programmes and allocation of
resources in Public Health practice

M105. THE MEGGA TRADING
COMPANY; UK
P. H. Hinings, Management Games Ltd;
Management Games Ltd
6th form and upwards
4-18; 6-8 hours
An interacting game on the operation of a
wholesale company buying materials for
resale in competitive situation; decisions
on buying and selling prices, advertising
expenditure, stock control and purchasing

M106. MEMORY; UK
—; Abbatt Toys Ltd
6-12 years old
Practice in memory training

M107. MEMORY GAME; UK
—; Waddingtons Ltd
3 years old and over
Any number
Practice in recognition of matching form and
colour

M108. MENTAL HEALTH; UK
—; Community Service Volunteers
Secondary school upwards
Introduction into a community of a hostel
for mentally subnormal young people

M109. MERGER; USA
—; Price Waterhouse & Co; Abt Associates Inc
College, management
Insight into how merger partners are chosen,
financial arrangements and the effects on
structure

M110. MERIT PICTURE DOMINOES;
UK
—; Merit Ltd
Primary school
Practice in shape and colour recognition

M111. MERIT PICTURE LOTTO;
UK
—; Merit Ltd
Primary school
1-7
Teaches picture and word matching

M112.   MERRY MILKMAN; 1972; UK
—; Merit Ltd
Primary school
2-4
Practice in counting

M113.   MERSEYSIDE YOUTH
        SERVICE TRAINING
        CENTRE; UK
—; Merseyside Training Centre
Youth leaders
Practice in youth service management

M114.   METFAB; 1969; USA
—; Macmillan Co; High School Geography
    Project
High school
5-10; 5 hours
Insight into determining optimum location
    for a manufacturing plant

M115.   METRO (MICHIGAN
        EFFECTUATION,
        TRAINING AND
        RESEARCH OPERATION);
        1966; USA
R. D. Duke; University of Michigan
College, town planning students
25-30
Training in urban planning

M116.   METRO-APEX; 1969; USA
R. D. Duke; University of Michigan, Environ-
    mental Simulations Laboratory
College, graduate school, management
9-39; 3-9 hours
Insight into municipal decision-making by
    politicians, industrialists, planners,
    developers and conservation officials

M117.   METROPOLIS; 1964; USA
R. D. Duke; University of Michigan,
    Environmental Simulations Laboratory
College, graduate school, management
3-45; 5-8 hours
Insight into municipal expenditure, urban
    growth and regional planning.

M118.   METROPOLITICS; 1971; USA
—; Western Behavioural Sciences Institute
Junior high school up to adults
18-35
Insight into problems and opportunities
    of various kinds of government

M119.   MICE TWICE; USA
P. Lamb; Lyons and Carnahan
Grades 1-6
3-4
Practice of long vowels; association with
    appropriate spelling patterns

M120.   MICRODIPLOMACY; UK
D. Miller; Albion
9
Diplomacy variant

M121.   MIDGARD; UK
H. Patterson; H. Patterson
General
Strategy and tactics game based on "sword
    and sorcery" science fiction

M122.   MIDTEX; 1969; USA
H. E. Johnson; College of Business
    Administration University of Texas
College, management
1+; ½ day
Teaches basic concepts of production and
    inventory control and Monte Carlo
    analysis

M123.   MIDVILLE HIGH SCHOOL
        PRINCIPALSHIP; 1965; USA
H. Laughlin; University Council for
    Educational Administration
Educational administrators
Practice conflict resolution

M124.   MIDWAY; 1964; USA
C. W. McClusky; Avalon-Hill Inc
General
2+; 2-6 hours
Insight into McClusky's strategy and tactics
    during the battle of Midway and practice
    in strategic decision-making

M125.   MILLEBORNES: UK
–; Parker Bros
10 years old
2-6
Practice French road traffic rules; travel in
France

M126.   MINI CUBE FUSION; UK
–; Waddingtons Ltd
General
2
Pocket version of 'Cube Fusion'

M127.   MINI-O-POLIS; USA
H. L. Barnett; H. L. Barnett
Junior high, high school
18-40; 3 hours +
Practice in community planning of a university
site

M128.   MISSION; 1969; USA
D. E. Yount, P. E. De Koch; Interact
Grade 7 to college
25-40; 5 hours
Insight into political positions in Vietnam
and South East Asia

M129.   MR. MONEY; 1972; UK
–; E. J. Arnold Ltd
Primary school
Practice in use of decimal currency

M130.   MR. PRESIDENT; USA
–; 3M Company
General
2-4; 1 hour
Insight into election campaign planning
and strategy

M131.   MR. PRESIDENT; 1971; USA
–; Wff'n Proof Inc
General
Identification of American Presidents from
range of clues, including misleading ones

M132.   MR. SPACE GAME; 1972; UK
–; E. J. Arnold Ltd
Primary school
Teaches maths terms

M133.   MR. TRUSTWORTHY'S
        FARM; UK
P. J. Wagland; St Mary's and St Paul's
Geographical School
1; ¾-1 hour
Insight into problems of an English farmer

M134.   M.I.T. MARKETING GAME;
        USA
–; Massachusetts Institute of Technology
College, management

M135.   M.I.T. POLITICAL-
        MILITARY EXERCISE;
        1960; USA
L. P. Bloomfield, B. Whaley; US Naval
Institute
Graduate school, military

M136.   MOCK-UP; UK
–; Bath Youth and Community Service;
Bristol District Methodist Association
Youth Clubs
Youth leaders
Practice leadership and decision-making

M137.   MOLYMOD ATOMS; UK
–; Spiring Enterprises
2-3
Construction of molecular models

M138.   MONAD; USA
S. Sackson; 3M Company
General
2-4
Strategic game on trading and buying

M139.   MONOPOLOGS; 1960; USA
J. R. Renshaw, A. Henston; Rand Corp
Military
1-30; 3 hours
Insight into inventory control problems of
an airline

M140.   MONOPOLY; 1935; USA
C. W. Darrow; Parker Bros
Primary, junior high, high school, general
2-8; 30 minutes +
Teaches economic concepts

M141.   THE MONROE CITY
        SUPERINTENDENCY; USA
—; University Council for Educational
    Administration
Educational administrators

M142.   MORDOR VERSUS THE
        WORLD IV
D. Miller *et al.*; Albion
General
5
A 'Diplomacy' variant based on 'Lord of the
    Rings'

M143.   MORE FAMILY PAIRS; UK
E. Root; Rupert Hart-Davis
Pre-readers
Teaches consonant blends

M144.   MORRA; Denmark
—; Piet Hein
General

M145.   MOSCOW CAMPAIGN; USA
—; Simulations Publications Inc
General
Tactical examination of the campaign

M146.   MOTIVATION; USA
—; Education Research
Management
Practice of appraisal interviewing in business

M147.   THE MOUSE IN THE
        MAZE; USA
—; Houghton Mifflin
Junior high, high school
1+
Demonstrates principles of learning
    behaviour and experimental design

M148.   MOVIEMAKER; USA
—; Parker Bros
General
Competitive board game based on film
    business

M149.   MULBERRY; 1971; USA
D. Dodge, *et al.*; Macalester College
High school and upwards
Insight into using federal resources and
    planning governmental structure

M150.   MULTICODE; UK
—; Griffin & George
Secondary school
Insight into processes of computers

M151.   MULTIPLICATION; USA
—; D. C. Heath Co; Abt Associates Inc
Elementary school
Practice in arithmetic drills

M152.   MULTIPLICATION BINGO;
        1971; UK
P. Holmes; Maths in School
Primary, secondary school
Practice in multiplication

M153.   MULTIPUZZLE; UK
—; Spears Games
10 years old to adult
Practice in manipulation of geometric shapes

M154.   MUSKET AND PIKE; USA
—; Simulations Publications Inc
General
War game

M155.   MY PICTURE WORD
        GAMES; 1972; UK
A. Stephenson; Galt Toys
5-8 years old
Practice in matching words and pictures

M156.   MY WORD BINGO; 1971; UK
N. Hindmarch; Educational Innovations
Pre-readers
2
Teaches that written words have meaning,
    that they are made up of letters and
    that each letter appears in a particular
    position in relation to other letters

M157.   MYTHIA: A WORLD
        AFFAIRS SIMULATION;
        1969; USA
J. M. Oswald, J. Pisano; American Institute
    for Research
Junior high, high school, college
8-40; 1 hour +
Increases perception of international
    endeavours; encourages co-operative
    attitudes to them

M158.   MYTHOMACY; UK
T. Kuch; Albion
General
Diplomacy variant

N1.   NAPOLEON AT BAY; USA
–; Strategy and Tactics
General
Napoleonic warfare

N2.   NAPOLEON AT WATERLOO;
USA
–; Simulations Publications Inc
General
1½ hours
Insight into Battle of Waterloo

N3.   NAPOLEON AT WATERLOO:
EXPANSION KIT; USA
–; Simulations Publications Inc
2 hours
Extension of 'Napoleon at Waterloo';
more complex rules and more counters

N4.   NAPOLI; 1966; USA
–; Western Behavioural Sciences Institute;
Simile II
Grade 6 up to graduate school
9-33; 2½ hours
Illustrates conflict between the party, the
constituency and the personal views of
the elected leglislator

N5.   THE NATIONAL ECONOMY;
1968; USA
E. Rausch, H. J. Cranmer; Science Research
Associates, Didactic Systems Inc
High school, college, management
3+; 2-5 hours
Insight into relationship of growth,
inflation, national income and
unemployment

N6.   NATIONAL ECONOMY GAME;
UK
I. Bracken; University of Wales, Cardiff
University
10+; 1 hour
Functioning of British national economy

N7.   NATIONAL MANAGEMENT
GAME (ANNUAL EVENT);
1970; UK
–; Financial Times; International Computers
Ltd; Institute of Chartered Accountants
Adults
3-5 teams (in final)
Insight into managing a firm producing and
marketing a consumer durable product

N8.   THE NATIONWIDE INSURANCE
MODEL; 1961; USA
R. T. Sampson; Nationwide Insurance
Management
5-40; 10-25 hours
Insight into whole-enterprise interactions of
functional decisions in a competetive
economy

N9.   NEGOTIATION; USA
–; Massachusetts Institute of Technology
University
Insight into processes of negotiation through
history of 1950s and relationship of
USA, USSR and UK

N10.   NEGOTIATION; USA
–; KDJ Instructional Systems Inc;
Educational Development Center
Grade 8
Insight into political compromise in the
setting of 1789 America

N11.   NEGOTIATIONS GAME; UK
N. Rackham; Industrial and Commercial
Training (June 1972 issue)
Management

N12.   NEIGHBORHOOD; USA
Abt Associates Inc; Wellesley School
System; Abt Associates Inc.
High school
Insight into growth of urban areas illustrated
by growth of Bostons' North End

N13.   NEMO (NUCLEAR EXCHANGE
MODEL); USA
–; Army Strategy & Tactics Analysis Group
Military
Insight into nuclear strategic war

N14. THE NESTING BOXES GAME;
1969; USA
—; Training Development Center
Management
4-16; 1-2 hours
Insight into planning, scheduling, directing
and controlling work

N15. NEUES KRIEGSSPIES; 1800;
Germany
—; Viturinus
Military
Tactical and logistic planning

N16. A NEW DISTRIBUTION
SYSTEM; UK
—; Esso
Secondary school to management
A study of the economics of distribution

N17. A NEW OIL; UK
—; Esso
Secondary school up to management
A study of market research and development

N18. NEW PENNY POCKET
MONEY; 1972; UK
—; E. J. Arnold Ltd
Primary school
2-4
Practice in use of decimal currency in
buying situation

N19. NEW TOWN: GAME 1;
1969; USA
B. R. Lawson; Harwell Associates
College, graduate school
2-10; 3-10 hours
Insight into planning and land development

N20. NEW TOWN: GAME 2;
1969; USA
B. R. Lawson; Harwell Associates
College, graduate school
2-10; 3-10 hours
Insight into planning and land development

N21. NEW TOWN: GAME 3;
1969; USA
B. R. Lawson; Harwell Associates
College, graduate school
2-10; 3-10 hours
Insight into planning and land development

N22. NEW TOWN: GAME 4;
1969; USA
B. R. Lawson; B. R. Lawson
College, graduate school
2-10; 3-10 hours
Insight into planning and land development

N23. NEW YORK STATE
REGIONAL HOUSING
MODEL; 1971; USA
D. W. Sears; University of Massachusetts
Public policy markets
Insight into problems of housing market

N24. NEWS (NAVY ELECTRONIC
WAR SIMULATOR); USA
—; US Naval War College
Naval
Insight into naval strategy and tactics

N25. NEWTON SPARKS AND
THE BOMB CHAIR
INCIDENT; 1970; UK
C. Adams; Centre for Structural Communication
Secondary school
Insight into electrical circuitry; practice in
completion of circuit diagrams

N26. THE NEXT PRESIDENT;
USA
—; Simulations Publications
General
Presidential election campaign in the US

N27.   NEXUS; 1970; UK
R. H. R. Armstrong, M. Hobson; Institute
of Local Government Studies,
Birmingham University
Post-experience, post-graduate, local
government
4+; 2 hours +
A 'frame technique' appropriate to many
situations; see also MANAGEMENT
BY OBJECTIVES, PROGRAMME
PLANNING SIMULATION,
CONTIGENCY RESOURCE
ALLOCATION and others. Demonstrates
and provides insight into the operation
of complex systems

N28.   NILE; UK
−; J. W. Spear & Sons Ltd
8 years and over
2-4
Teaches tactical thinking

N29. NIMBI; 1972; Denmark
Piet Hein; Piet Hein
General
A 12 counter strategic game on a circular
board

N30.   NI PRACTICE; 1962; USA
L. E. Allen; Wff'n Proof Inc
Grade 1 to college
2+
Teaches use of negation-in rule of logic

N31.   1914 (GAMES 1, 2, 3, 4,
5, 6 AND 7); 1968; USA
−; Avalon-Hill Inc
General
2+; 4-10 hours
Insight into First World War

N32.   1918; USA
−; Simulations Publications Inc
General
Insight into 1918 Western front:

N33.   1925 TO 1975 INFANTRY
ACTION; 1971; UK
Wargames Research Group; Wargames
Research
General
Insight into tactics and logistics of modern
warfare

N34.   1940: THE BATTLE FOR
FRANCE; UK
−; Strategy And Tactics
General
German invasion of France

N35.   THE 1972 WAR IN
THAILAND; 1965; USA
W. & J. Fain; Douglas Aircraft Corp
Adults, military
Insight into insurgency in Thailand

N36.   NO DAM ACTION; USA
−; Instructional Simulations Inc
Grades 8-12
Insight into human ecology

N37.   NO NONSENSE; USA
P. Lamb; Lyons & Carnahan
Grades 1-6
4-8
Reinforces morphemic principles

N36.   NO PRACTICE; 1962; USA
L. E. Allen; Wff'n Proof Inc
Grade 1 to college
2+
Teaches use of negation-out rule of logic

N39.   NOMINATING CONVENTION
GAME; USA
S. A. Schainker; Horton Watkins High
School
High school
US Presidential elections

N40.   NORMANDY; USA
−; Simulations Publications Inc
General
Normandy landings in 1944

N41. THE NORTHEAST FARM
MANAGEMENT GAME
(FMG 4); 1968; USA
E. I. Fuller; Department of Agriculture
and Food Economics, University of
Massachusetts
College, management
1-100; 3-7 hours
Creates empathy with and gives practice
in heuristic and analytic management
decision-making

N42. NORTH SEA EXPLORATION;
1973; UK
R. Walford; Longmans
Secondary school
Up to 30; 3 hours
Economics, geography etc of exploration
for North Sea gas and oil

N43. THE NORTH SEA GAS
GAME; 1969; UK
R. Walford; Longmans Green
Secondary school
Insight into the problems of mineral search
companies, North Sea Gas exploitation
and practice in search techniques

N44. NUCLEAR DESTRUCTION;
1972; USA
R. Loomis; R. Loomis
General
10-12
War game played on a computer

N45. NUCLEAR WAR; 1967; USA
-; Simulations Publications Inc
General
Strategy and tactics

N46. NUMBER AND PICTURE
DOMINOES; UK
-; E. J. Arnold Ltd
Primary school
Group and symbol matching

N47. NUMBER DOMINOES: UK
-; Abbatt Toys Ltd
Primary school
Group and symbol number matching

N48. NUMBER LOTTO; 1972; UK
-; E. J. Arnold
Primary school
Practice in number recognition

N49. NUMBERS UP; UK
Waddingtons Ltd
8 years old and over
2+
Teaches number and four rules

N50. NUMERO; 1972; UK
-; Games and Puzzles
Children and upwards
2
Maths roots

O1. OASIS (OFFICE
ADMINISTRATION
SIMULATION STUDY);
1968; USA
B. Shorr, P. Hagstrom, J. Davis, L. Buck;
Travelers Insurance Company
Management
8-40; 8 days
Familiarity with a new variable budget system

O2. OBJECT PUZZLE DOMINOES;
UK
-; Abbatt Toys Ltd
4-8 years
Shape recognition

O3. ODYSSEY; USA
-; Magnavox
General
2
Electronic (TV) games of football etc

O4. OFFICE SIMULATION;
1971; USA
M. J. Krawitz; McGraw-Hill
High school
6-30; 9-18 weeks
Practice in application of office skills

O5. OH WAH REE; USA
-; 3M Company
General
Tactical board game

O6. THE OIL EXPLORATION
GAME; 1972; UK
J. P. Cole, P. M. Mather, P. T. Whyrall;
Department of Geography, Nottingham
University
Secondary school upwards
1
Operational model of off-shore exploration
and extraction of oil and marketing
system. Can be played on a computer,
also teaches programming skills

O7. OKLAHOMA BUSINESS
MANAGEMENT GAME; USA
B. Sanders; Oklahoma State University
College
30-50; 20 hours
Insight into way management teams operate
and into inter-relations between teams

O8. THE OKLAHOMA
PETROLEUM MANAGEMENT
GAME; USA
E. Z. Million; Computer Congenerics Corp
Management
15-40; 1 hour
Insight into problems and decision-making
in functional areas other than participants
own

O9. OLD ITCH; USA
A. W. Heilman, R. Helmkamp, A. E. Thomas,
C. J. Carsello; Lyons and Carnahan
Grades 1-3
2+
Practice in auditory perception of initial
consonants

O10. ON SETS: A–B; 1969; USA
L. E. Allen, I. P. Kugel, M. Owen; Wff'n
Proof Inc; Autotelic Instructional
Material Publishers
Young children
1
Practice in devising set names that match
given goals

O11. ON SETS: ADVANCED;
1969; USA
L. E. Allen, I. P. Kugel, M. Owen; Wff'n
Proof Inc; Autotelic Instructional
Material Publishers
Adults
2+
Practice of and insight into advanced set
theory

O12. ON SETS: BASIC; 1969; USA
L. E. Allen, I. P. Kugel, M. Owen; Wff'n
Proof Inc, Autotelic Instructional Material
Publishers
Young children
2+
Insight into and practice with elementary
set theory

O13. ON SETS: CAP FISH; 1969;
USA
L. E. Allen, I. P. Kugel, M. Owen; Wff'n
Proof Inc, Autotelic Instructional
Material Publishers
Young children
2+
Practice in use of symbol ∩

O14. ON SETS: CAP SETS; 1969;
USA
L. E. Allen, I. P. Kugel, M. Owen; Wff'n
Proof Inc, Autotelic Instructional
Material Publishers
Young children
2+
Prepares for 'On-Sets' and gives practice in
use of symbol ∩

O15. ON SETS: CAP SQUAD;
1969; USA
L. E. Allen, I. P. Kugel, M. Owen; Wff'n
Proof Inc, Autotelic Instructional
Material Publishers
Young children
2+
Practice in use of symbol ∩

O16. ON SETS: COMP FISH; 1969;
USA
L. E. Allen, I. P. Kugel, M. Owen; Wff'n
Proof Inc, Autotelic Instructional
Material Publishers
Young children
2+
Practice in use of symbol /

O17. ON SETS: COMP SETS;
1969; USA
L. E. Allen, I. P. Kugel, M. Owen; Wff'n
Proof Inc, Autotelic Instructional
Material Publishers
Young children
2+
Practice in use of symbol / and preparation
for 'On-Sets'

O18. ON SETS: COMP SQUAD;
1969; USA
L. E. Allen, I. P. Kugel, M. Owen; Wff'n
Proof Inc, Autotelic Instructional
Material Publishers
Young children
2+
Practice in use of symbol /

O19. ON SETS: CUBE FISH; 1969;
USA
L. E. Allen, I. P. Kugel, M. Owen; Wff'n
Proof Inc, Autotelic Instructional
Material Publishers
Young children
2+
Teaches concept of set by colour

O20. ON SETS: CUBE SETS; 1969;
USA
L. E. Allen, I. P. Kugel, M. Owen; Wff'n
Proof Inc, Autotelic Instructional
Material Publishers
Young children
2+
Teaches basic structure of 'On-Sets' and
gives practice in the concept of set

O21. ON SETS: CUBE SQUAD;
1969; USA
L. E. Allen, I. P. Kugel, M. Owen; Wff'n
Proof Inc, Autotelic Instructional
Material Publishers
Young children
2+
Teaches concept of set defined by property

O22. ON SETS: CUP FISH; 1969;
USA
L. E. Allen, I. P. Kugel, M. Owen; Wff'n
Proof Inc, Autotelic Instructional
Material Publishers
Young children
2+
Practice in use of symbol $\cup$

O23. ON SETS: CUP SETS; 1969;
USA
L. E. Allen, I. P. Kugel, M. Owen; Wff'n
Proof Inc, Autotelic Instructional
Material Publishers
Young children
2+
Introduction to new features of 'On-Sets'
and practice in the use of symbol $\cup$

O24. ON SETS: CUP SQUAD;
1969; USA
L. E. Allen, I. P. Kugel, M. Owen; Wff'n
Proof Inc, Autotelic Instructional
Material Publishers
Young children
2+
Practice in use of symbol $\cup$

O25. ON SETS: DIFF FISH; 1969;
USA
L. E. Allen, I. P. Kugel, M. Owen; Wff'n
Proof Inc, Autotelic Instructional
Material Publishers
Young children
2+
Practice in use of symbol $-$

O26.   ON SETS: DIFF SETS; 1969;
       USA
L. E. Allen, I. P. Kugel, M. Owen; Wff'n
   Proof Inc, Autotelic Instructional
   Material Publishers
Young children
Practice in use of symbol – and preparation
   for 'On-Sets'

O27.   ON SETS: DIFF SQUAD;
       1969; USA
L. E. Allen, I. P. Kugel, M. Owen; Wff'n
   Proof Inc, Autotelic Instructional
   Material Publishers
Young children
Practise with the symbol –

O28.   ON SETS: EQ FISH; 1969;
       USA
L. E. Allen, I. P. Kugel, M. Owen; Wff'n
   Proof Inc, Autotelic Instructional
   Material Publishers
Young children
2+
Practice in use of symbol =

O29.   ON SETS: EQ OUT; 1969;
       USA
L. E. Allen, I. P. Kugel, M. Owen; Wff'n
   Proof Inc, Autotelic Instructional
   Material Publishers
Young children
2
Practice in use of symbol =

O30.   ON SETS: EQ SETS; 1969;
       USA
L. E. Allen, I. P. Kugel, M. Owen; Wff'n
   Proof Inc, Autotelic Instructional
   Material Publishers
Young children
2+
Practice in use of symbol = and preparation
   for 'On-Sets'

O31.   ON SETS: EQ SQUAD; 1969;
       USA
L. E. Allen, I. P. Kugel, M. Owen; Wff'n
   Proof Inc, Autotelic Instructional
   Material Publishers
Young children
2+
Practice in use of symbol =

O32.   ON SETS: GO THROUGH;
       1969; USA
L. E. Allen, I. P. Kugel, M. Owen; Wff'n
   Proof Inc, Autotelic Instructional
   Material Publishers
Young children
2+
Set matching practice

O33.   ON SETS: INC FISH; 1969;
       USA
L. E. Allen, I. P. Kugel, M. Owen; Wff'n
   Proof Inc, Autotelic Instructional
   Material Publishers
Young children
2+
Practice in use of symbol $\subseteq$

O34.   ON SETS: INC OUT; 1969;
       USA
L. E. Allen, I. P. Kugel, M. Owen; Wff'n
   Proof Inc, Autotelic Instructional
   Material Publishers
Young children
2
Practice in use of symbol $\subseteq$

O35.   ON SETS: INC SETS; 1969;
       USA
L. E. Allen, I. P. Kugel, M. Owen; Wff'n
   Proof Inc, Autotelic Instructional
   Material Publishers
Young children
2+
Practice in use of symbol $\subseteq$ and preparation
   for 'On-Sets'

O36. ON SETS: INC SQUAD;
1969; USA
L. E. Allen, I. P. Kugel, M. Owen; Wff'n
Proof Inc, Autotelic Instructional
Material Publishers
Young children
2+
Practice in use of symbol ⊆

O37. ON SETS: ONE; 1969; USA
L. E. Allen, I. P. Kugel, M. Owen; Wff'n
Proof Inc, Autotelic Instructional
Material Publishers
Young children
1
Practice in solving problem of finding set
with one card

O38. ON SETS: OUT CAP; 1969;
USA
L. E. Allen, I. P. Kugel, M. Owen; Wff'n
Proof Inc, Autotelic Instructional
Material Publishers
Young children
2
Practice in use of symbol ∩

O39. ON SETS: OUT COMP; 1969;
USA
L. E. Allen, I. P. Kugel, M. Owen; Wff'n
Proof Inc, Autotelic Instructional
Material Publishers
Young children
2
Practice in use of symbol

O40. ON SETS: OUT CUBE; 1969;
USA
L. E. Allen, I. P. Kugel, M. Owen; Wff'n
Proof Inc, Autotelic Instructional
Material Publishers
Young children
2
Teaches concept of set by property

O41. ON SETS: OUT CUP; 1969;
USA
L. E. Allen, I. P. Kugel, M. Owen; Wff'n
Proof Inc, Autotelic Instructional
Material Publishers
Young children
2
Practice in use of symbol ∪

O42. ON SETS: OUT DIFF; 1969;
USA
L. E. Allen, I. P. Kugel, M. Owen; Wff'n
Proof Inc, Autotelic Instructional
Material Publishers
Young children
2
Practice in use of symbol −

O43. ON THE BUTTON; USA
−; Abt Associates Inc
Elementary school
2
Arithmetic drills

O44. ON-WORDS; 1971; USA
−; Wff'n Proof Inc
Primary school
Practice in spelling and word combinational
problem-solving

O45. 1000 BC TO AD 1000; 1969;
UK
−; Wargames Research Group
General
Insight into tactics of ancient warfare

O46. ONE + ONE = TWO; Holland
−; Jumbo Games
Primary school
Addition and subtraction

O47. ONE TO THREE GAME;
1967; USA
H. A. Springle; Science Research Associates
Pre-school to grade 1
1-4; 20-30 minutes
Teaches recognition of numerals and the
sets they represent

O48. ONE TO SIX GAME; 1967; USA
H. A. Springle; Science Research Associates
Pre-school to grade 1
1-4; 20-30 minutes
Teaches recognition of numerals and the
sets they represent

O49. ONE TO NINE GAME; 1967; USA
H. A. Springle; Science Research Associates
Pre-school to grade 1
1-4; 20-30 minutes
Teaches recognition of numerals and the
sets they represent

O50. OPERATION CAPRICE; USA
—; Harvard University
Arts administrators
Decision-making and resource allocation in
context of management of an arts centre

O51. OPERATION GREIF; USA
—; Guidon Games
General
Ardennes campaign 1944

O52. OPERATION OF DISTRICT
YOUTH SERVICE
STRUCTURE; UK
A. Aldrich; Wiltshire Training Agency
Youth leaders
Practice in decisive leadership and
co-operation for full-time youth service
staff; insight into the need for
organisation

O53. OPERATION OF A LEADER-
SHIP TRAINING CENTRE;
UK
A. Aldrich; Wiltshire Training Agency
Youth leaders
Practice in decisive leadership and
co-operation for part-time leaders;
insight into the need for organisation

O54. OPERATION SUBURBIA;
1961; USA
A. A. Zoll; Addison-Wesley
Management
1 hour
Insight into town planning

O55. OPERATION TAURUS; 1962;
UK
Webb, Wheeler; Howard Farrow Ltd
Management
Insight into economics of building and
civil engineering firms

O56. OPSIM; 1969; USA
B. R. Darden, W. H. Lucas; Appleton-
Century-Crofts
Management
An integrated operations management game
based on the management of a production
department

O57. ORIGINS OF WORLD
WAR II; USA
—; Avalon-Hill Inc
General
2-5
Insight into diplomatic conflicts of
1935-1939

O58. ORION; USA
—; Parker Bros
General
Strategy game in space

O59. ORWELL SCHOOL; UK
—; Berkshire College of Education
Student teachers
Practice in decision making in school
context

O60. OTTO; Germany
—; OKH
Military
Insight into problems of an invasion of
Russia

O61. OUTDOOR SURVIVAL; USA
J. Dunnigan; Avalon-Hill Inc
Teaches survival techniques

P1 PACIFIC EXPRESS; USA
—; University of Michigan
High school
Insight into the construction of trans-
continental railways; practice in decision-
making

P2. PACIFIC '42; USA
—; Simulations Publications Inc
General
World War II

P3. PAIR-IT; UK
E. Root; Rupert Hart-Davis Ltd
Pre-readers
Teaches word recognition

P4. PAIRING; 1937; UK
F. J. Schonell; Oliver & Boyd Ltd
Primary school, remedial classes
3
Practise addition

P5. PANATINA; USA
J. Reese; J. Reese
Grades 5 and 6
15-30; ½-1 hour
Insight into the problems of a northern
Andes state poor in resources but
potentially rich

P6. PANATINA; USA
—; Project Simile
High school
18-35; 5-6 hours
Problems of land reform, revolution and
common markets in South America

P7. PANIC; 1968; USA
D. Yount, P. De Koch; Interact
Grade 7 up to college
25-36; 23 days
Insight into the economics and political
pressure groups of the 1920's in the US

P8. PANZERBLITZ; USA
—; Avalon-Hill Inc
General
2-5 hours
Practice in tactical thinking in context of
World War II Russian front

P9. PARA-TIME DIPLOMACY; UK
D. Miller; Albion
General
Diplomacy variant

P10. PARENT-CHILD GAME;
1969; USA
E. O. Schild, S. S. Boocock; John Hopkins
University
High school
4-10; 1 hour
Insight into parent/child relationship and
permissible behaviour

P11. PARLEMENT: USA
C. Wells; Albion
Insight into working of the French political
system

P12. PARTICIPATIVE DECISION-
MAKING; 1970; USA
R. E. Horn; Information Resources Inc
High school, college
6+; 2-4 hours
Gives experience of gaming; teaches basic
concepts of participative decision-making
and programme budgeting

P13. PATCH MATCH; USA
P. Lamb; Lyons & Carnahan
Grades 1-6
2-8
Teaches beginning and ending consonant and
short vowel sounds, graphemes based on
these and graphemic applications

P14. PATIENCE; 1958; UK
L. W. Downes, D. Paling; Oxford University
Press
Primary school
1
Practice in multiplication

P15. PATROL; USA
—; Simulations Publications Inc
General
Insight into tactics in Vietnam war

P16. PAY THE CASHIER; 1957;
USA
E. W. Dolch; Garrard Publishing Co
Kindergarten to Grade 3
2-6; 10-15 minutes
Teaches currency up to $10, its addition,
making change

P17.   PC PROOF; 1962; USA
L. E. Allen; Wff'n Proof Inc
Grade 1 up to college
2+
Practice in use of propositional calculus and
   derived rules of logic

P18.   PENALTY; Italy
E. Scola; Pepys Games
General
2-6
Football

P19.   PENALTY; UK
—; Castell Ltd
General
Football

P20.   THE PENTAGON GAME;
        1972; UK
B. Beckett, K. Cuthbertson; Infinity
   Communications Ltd
General
2-4; 1-3 hours
Vietnam war; model of counter-insurgency
   warfare

P21.   PERCEPTION GAMES; 1970;
        USA
H. Hoolim; MASCO
Infants
1-8
Practice in perception

P22.   PERCEPTUAL LEARNING
        PUZZLES–BALLOONS;
        1969; USA
L. G. Gotkin; New Century
Primary school
1; 20-30 minutes
Discrimination of sizes of circle and various
   fractions.

P23.   PERCEPTUAL LEARNING
        PUZZLES–CLOWNS; 1969; USA
L. G. Gotkin; New Century
Primary school
1; 20-30 minutes
Practice in discrimination of basic alphabet
   shapes

P24.   PERCEPTUAL LEARNING
        PUZZLES–LANGUAGE
        SKILLS; 1969; USA
L. G. Gotkin; New Century
Primary school
1; 20-30 minutes
Practice in use and eliciting vocabulary

P25.   PERCEPTUAL LEARNING
        PUZZLES–VISUAL
        DISCRIMINATION 1; 1969;
        USA
L. G. Gotkin; New Century
Primary school
1; 20-30 minutes
Teaches size and shape discrimination

P26.   PERCEPTUAL LEARNING
        PUZZLES–VISUAL
        DISCRIMINATION 2; 1969;
        USA
L. G. Gotkin; New Century
Primary school
1; 20-30 minutes
Teaches mirror symmetry and detail
   discrimination

P27.   THE PERFORMANCE GAME;
        1969; USA
—; Training Development Center
Management
6-15; 1-2 hours
Practice in evaluating, increasing and
   rewarding performance of employees

P28.   A PERSONAL ASSESSMENT;
        UK
—; Careers Research Advisory Centre
Secondary school, college, adults
Exercise on job analysis and enrichment

P29.   PERSONALYSIS; 1965; USA
Administrative Research Association
High school, college, management
3-4; 1 hour
Insight into own personality; practice in
   adaption and improvement of relationships

P30. PERSONNEL; 1968; USA
E. Rausch; Science Research Associates
Management
6+; 2-5 classes
Testing of theory in problem-solving

P31. PERSONNEL ASSIGNMENT
MANAGEMENT GAME;
USA
J. R. Greene, R. L. Sisson; Didactic Systems
Inc, J. Wiley
Management
1-2
Insight into a simple linear programming
problem

P32. THE PERSONNEL
DEPARTMENT; USA
J. J. Zif, A. H. Walker, E. Orbach;
MacMillan & Co
College management
20-40; 2½-6 hours
Insight into functions and issues of personnel
management, experience in conflict
resolution and group decision making
practice in application of personnel
theory to real life

P33. PERSONNEL INTERVIEWING;
USA
E. Rausch; Science Research Associates
Management.
2+

P34. PERSONNEL MANAGE-
MENT EXERCISE; 1965; USA
W. Howells, J. N. Fairhead, D. S. Pugh,
W. J. Williams; E.U.P.
Management
Insight into process of optimization of
staffing ratios and stability to ensure
adequate ability to grow rapidly

P35. PERTSIM; 1969; USA
L. A. Swanson, H. L. Pazer; Didactic
Systems Inc; International Textbook Co
College, management
3+; 5 days
Demonstrates fundamentals of PERT and
CPM in construction management

P36. PETER RABBIT RACE
GAME; UK
—; Frederick Warne Ltd
5-8 years old

P37. PHALANX; USA
—; Simulations Publications Inc, Whitman
Publishing Co
General
Insight into tactics of Greek, Roman, Persian,
Spartan and Macedonian armies
500-100 B.C.

P38. PHYSICAL DISTRIBUTION
MANAGEMENT; USA
—; Didactic Systems Inc
Management
3-5
Insight into responsibilities of physical
distribution

P39. PICKAFIT; USA
P. McKee, M. L. Harrison; Houghton
Mifflin, Abt Associates Inc
Pre-reading
1
Practice in using context with beginning
consonant sound and consonant letter

P40. PICTURE BINGO; 1970; UK
—; Community Service Volunteers
Immigrants
3+
Phonic practice

P41. PICTURE DOMINOES; UK
—; Abbatt Toys Ltd
7 years old
Practice in picture matching

P42. PICTURE DOMINOES; UK
—; Spears' Games
4 years old and over
Practice in relating pictures to names

P43. PICTURE LOTTO; UK
F. Lewis; Galt Toys
4-5 years old
2-6
Insight into keeping to rules

P44. PICTURE PUZZLE CLOCK;
UK
–; Victory Ltd
Primary school
1
Practice in telling the time

P45. PICTURE READINESS; 1949;
USA
E. W. Dolch; Garrard Publishing Co
Primary school
1-6; 10-15 minutes
Promotes attentiveness and perception of
details

P46. PICTURE WORDS; USA
P. McKee; M. L. Harrison; Houghton
Mifflin; Abt Associates Inc
Pre-reading
1
Practice in using context and sounds of
beginning, middle and ending consonants
as clues to words unfamiliar in printed form

P47. THE PIES GAME; 1972; UK
J. P. Cole, P. M. Mather, P. T. Whysall;
Department of Geography, Nottingham
University
Secondary school upwards
Pie manufacturers compete for markets;
game involves pricing policy, transport
costs and shows importance of business
location. Can be played on computer,
also teaches programming skills

P48. PLANAFAM; USA
K. Finseth; Social Education
General
Insight into planning and decision-making in
US Indian family on the basis of values,
traditions and circumstances

P49. PLANDEC; 1971; USA
J. Pfeffer, H. R. Fogler, T. Deeley;
Stanford University
Management
Insight into strategic and tactical planning
in management finance and into data
processing

P50. THE PLANET MANAGE-
MENT GAME; USA
–; Houghton Mifflin
Junior high school, college
2-12
Demonstrates results of decision-making in
economic and environmental change

P51. PLANNED MAINTENANCE;
1969; USA
–; Didactic Systems Inc
Management trainees
3-5; 2-3½ hours
Practice problem-solving and maintenance
management skills

P52. PLANNING, PROGRAMMING,
BUDGETING SYSTEM–
EDUCATION: COST UTILITY
ANALYSIS; 1970; USA
V. C. Gideon; Information Resources Inc
Management, educators
1 hour
Orientation to PPBS and practice in
analysing alternatives in terms of costs

P53. PLANNING, PROGRAMMING,
BUDGETING SYSTEM–
EDUCATION: DEFENDING
THE BUDGET; 1970; USA
V. C. Gideon; Information Resources Inc
Management, educators
1 hour
Orientation to PPBS and defending decisions
thus made

P54. PLANNING, PROGRAMMING,
BUDGETING SYSTEM–
EDUCATION: DEVELOPING
PROGRAMME STRUCTURES;
1970; USA
V. C. Gideon; Information Resources Inc
Management, educators
45 minutes
Orientation to PPBS, practice in developing
syllabi

P55. PLANNING, PROGRAMMING, BUDGETING SYSTEM— EDUCATION: EXAMINING PROGRAMME STRUCTURES; 1970; USA
V. C. Gideon; Information Resources Inc
Management, educators
1 hour
Orientation to PPBS, practice in critical assessment of syllabi

P56. PLANNING, PROGRAMMING, BUDGETING SYSTEM— EDUCATION: FORMULATING AND STATING EDUCATIONAL PHILOSOPHY, GOALS AND OBJECTIVES; 1970; USA
V. C. Gideon; Information Resources Inc
Management, educators
1 hour
Practice in stating objectives and aims

P57. PLANNING, PROGRAMMING, BUDGETING SYSTEM— EDUCATION: ISSUE PAPER; 1970; USA
V. C. Gideon; Information Resources Inc
Management, educators
1 hour
Orientation to PPBS, practice in planning issue papers

P58. PLANNING, PROGRAMMING, BUDGETING SYSTEM— EDUCATION: PROGRAMME MEMORANDUM; 1970; USA
V. C. Gideon; Information Resources Inc
Management, educators
45 minutes
Practice in examination of programme (syllabus) memorandum to determine deficiencies

P59. PLANNING, PROGRAMMING, BUDGETING SYSTEM— EDUCATION: SYSTEMS ANALYSIS; 1970; USA
V. C. Gideon; Information Resources Inc
Management, educators
20 minutes
Practice in analysing and modifying a systems analysis report by consultants

P60. PLANNING, PROGRAMMING, BUDGETING SYSTEM— EDUCATION: TRAINING SYSTEMS ANALYSIS; 1970; USA
R. E. Horn; Information Resources Inc
Management, educators
1 hour
Practice in examination of who needs which kind of training and the production of a draft schedule

P61. PLANNING THE ADVER- TISING CAMPAIGN; 1970; USA
J. R. G. Jenkins, J. J. Zif; McMillan
College, management
9-18; 4-5 hours
Practice in planning an advertising campaign in a logical way; increases familiarity with advertising activities and responsibilities

P62. PLANS; 1966; USA
R. Boguslaw, R. H. Davis, E. B. Crick; Western Behavioural Sciences Institute, Simile II, Systems Development Corp
High school, college, graduate school
12-36; 100-300 minutes
Insight into actions and consequences of pressure groups in national politics

P63. PLATO; 1972; UK
—; Habitat
5 years old

P64. PLAYING CARD NUMBER GAME; USA
—; D. C. Heath & Co
Primary school
Practice in operations with whole numbers

P65. PLAY ON WORDS; UK
—; Waddingtons Ltd
8 years old and over
2+
Teaches strategic thinking, spelling and vocabulary

P66.   PLAYOFF COMPUTERIZED
       BASKETBALL; USA
—; E. S. Lowe
General
2

P67.   PLOY; USA
—; 3M Company
General
2-4; 30 minutes
Tactical board game

P68.   P.O.G.E. (PLANNING,
       OPERATIONAL GAMING
       EXPERIMENT); 1960; USA
Hendricks; American Institute of Planners
Management
Management planning and strategy

P69.   POINT ROBERTS; 1969; USA
—; McMillan Co, High School Geography
       Project
High school
Up to 30; 6 days
Insight into problems arising from historical
   events; practice in working with
   alternative solutions to an international
   boundary dispute

P70.   POLICY NEGOTIATIONS; USA
F. Goodman; University of Michigan
College of Education
Insight into negotiations between teachers
   and a school board

P71.   POLIS (POLITICAL
       INSTITUTIONS
       SIMULATION); 1969; USA
R. C. Noël; University of California,
   American Behavioural Scientist
University
Insight into political community

P72.   POLITICA; USA
M. S. Gordon, D. DelSolar; Advanced
   Research Projects Agency, Abt
   Associates Inc
Military
Insight into the dynamics of prerevolutionary
   crises in Latin America

P73.   POLITICAL-MILITARY
       EXERCISE (PME); 1958; USA
—; Center for International Studies
   Massachusetts Institute of Technology
College, military
8-50; 1-5 days
Insight into complexities of real decision-
   making in foreign policy

P74.   POLITICAL PARTY
       NOMINATING GAME; USA
—; Horton Watkins High School
High school
American elections

P75.   POLITICS IN BENIN; USA
—; Edcom Systems Inc, Abt Associates
Grade 4
Illustrates the complex political system
   of Benin (Nigeria)

P76.   POLLUTION; 1970; USA
Abt Associates Inc; Abt Associates Inc
Primary school
15-25; 3¾-11¼ hours
Insight into social, political and economic
   problems of pollution control

P77.   POLLUTION; USA
P. A. Twelker; Instructional Development
   Corp
Grades 7-12 to adults
4-32
Insight into the relationship of personal or
   group goals and the quality of life, the
   reasons for the cost and complexity of
   pollution control and the strategies
   consistent with goals. Creates positive
   attitudes over pollution

P78.   POLLUTION; USA
F. A. Rasmussen; Houghton Mifflin Co
Junior high school to adults
4-5
Insight into bargaining skills and the role
   of the profit motive

P79. POLYCHOC; UK
Management Games Ltd; Management
Games Ltd·
6th form and upwards
1-24; 6-10 hours
Insight into problems of production and
marketing of a consumable with a
limited shelf-life; decisions on production
rate, quality and costs, demand forcasts,
trade and consumer advertising, salesman
incentives, market research

P80. POLYOPTOMY; 1971; UK
A. Ighodaro, M. Ford; Think Games Ltd
Adults
2
Executive desk game

P81. POLYTAIRE; Denmark
−; Piet Hein
General

P82. POLYTAIRE; USA
−; Didactic Systems Inc
General
5-8
Insight into group processes

P83. POPULATION; USA
−; Urban Systems Inc
12 years old to adults
Insight into the crisis of over-population
in a rapidly developing country

P84. PORTSVILLE; 1969; USA
Rutgers University; Macmillan Co., High
School Geography Project
High school
30; 500 minutes
Insight into the growth of an American city

P85. POTLACH; USA
−; Abt Associates Inc
Junior high school
Insight into the social and economic
institutions of the Kwakiutl Indians
(N.W. Pacific coast)

**P86. THE POULTRY FARM
MANAGEMENT GAME
(POULT 4); 1968; USA**
E. I. Fuller; Department of Agricultural
and Food Economics, University of
Massachusetts
College
1+; 3-7 hours
Creation of empathy for and practice of
heuristic and analytical management
decision-making

P87. POWDER HORN: USA
G. Shirts; Project Simile
High school, college
18-35; 1 hour +
Insight into three-tiered society, trading
rifles, traps and pelts

P88. POWER POLITICS; USA
R. Durham, V. Durham, P. A. Twelker;
Instructional Development Corporation
Grades 9-12 to adults
20-40
Practice in comprehension, interpretation,
analysis, synthesis and evaluation in
relation to politics

P89. PREDATOR PREY; USA
−; Urban Systems Inc
High school
Insight into the ecology of animal
competition and survival

P90. PREDICAMENTS; 1972; UK
A. Allkins; Games & Puzzles
Adults
Groups
Practice in logical and deductive thought

**P91. PREDICTION OF VOTING IN
THE UNITED NATIONS;
1971; USA**
D. Dodge; Macalester College
College
Insight into the processes of predicting
voting by other nations in the UN

P92. PREPARING FOR CHANGE:
UK
—; International Chemical Industries;
S. B. Modules Ltd
Supervisory staff
Insight into problems of explaining a
productivity deal

P93. PRESIDENTIAL ELECTION
CAMPAIGN; USA
L. Stitelman, W. D. Coplin; Science
Research Associates
High school, college
US elections

P94. PRESIDENTIAL ELECTION
CAMPAIGN IN WELBOLAND
(FULL VERSION); 1971; UK
J. P. Cole; Geography Department,
Nottingham University
Secondary school upwards
3
Planning of election campaign and speech-
making tour through constituencies

P95. PRESIDENTIAL ELECTION
CAMPAIGN IN WELBOLAND
(SHORT VERSION); 1971; UK
J. P. Cole; Geography Department, University
of Nottingham
Secondary school upwards
1
Planning of election campaign and speech-
making tour through constituencies

P96. PRIMARY NUMBER LOTTO;
1972; UK
—; E. J. Arnold Ltd
Primary school
Teaches colour recognition and number
combinations

P97. PRINCE: PROGRAMMED
INTERNATIONAL COMPUTER
ENVIRONMENT; 1971; USA
W. Coplin, M. K. O'Leary, S. Mills; Inter-
national Relations Programme, Syracuse
University
College
Insight into US foreign policy making

P98. PRIORITY; USA
—; Education Research
High school, college
Practice in time allocation for pay-off in
finance

P99. PRIORITY GAME; 1968; UK
D. F. Sutton; Intertext Ltd
Management trainees
Practice in arranging priorities for action

P100. PROBE; UK
—; Parker Bros
10 years old and over
2-4
Vocabulary practice

P101. PROBLEM GAME; 1972; UK
J. Hardy; Birmingham University
Post-graduate social workers
Practice in solving the variety of problems
facing a local authority department of
social services

P102. PROBLEM-SOLVING IN A
HIERARCHY; UK
—; Management Associates
College, management
Problem-solving

P103. PROBLEMS IN BANK
MANAGEMENT: AN
IN-BASKET TRAINING
EXERCISE; 1969; USA
—; Addison-Weslay Publishing Co
College, management
1; 1-2 hours
Practice in organizing, planning, perception,
leadership and sensitivity

P104. PROBLEMS IN SUPER-
VISION (AN IN-BASKET
EXERCISE); 1968; USA
C. L. Jaffee; Addison Wesley Publishing Co
College, management
1; 1-2 hours
Practice in skills of organizing, planning,
perception, leadership and sensitivity

P105. PRODUCT PLAN; 1966; USA
M. Hanan; Hanan & Son
Management
15-45; 7 hours
Practice in creation and management of new
products

P106. PRODUCTION CONTROL
INVENTORY; 1968; USA
E. Rausch; Didactic Systems Inc, Science
Research Associates
Management students
5+; 2½ hours
Practice in and exchange of information on
approaches to good inventory control
management

P107. PRODUCTION CONTROL
SCHEDULING; 1968; USA
E. Rausch; Science Research Associates
Management
6+; 2-5 hours
Practice in testing problem-solving theory

P108. PRODUCTION PLANNING
EXERCISE; 1965; UK
B. Gould, J. N. Fairhead, D. S. Pugh,
W. J. Williams; E.U.P.
Management
Insight into the need for programming
techniques in production industry

P109. PRODUCTION SCHEDULING
MANAGEMENT GAME; USA
—; Didactic Systems Inc; J. Wiley
Management trainees
1
Practice in scheduling of production
facilities

P110. PRODUCTION SIMULATION
PROJECT; Canada
P. R. Winters; Faculty of Business,
University of Alberta
College
1+; 1-2 months
Provides background knowledge for project
on systems analysis and design

P111. PROFESSIONAL
COLLABORATION
EXERCISE; 1967; UK
J. E. Cooke; Department of the
Environment
Management
2¾ days
Improves collaboration between professionals
in the design process

P112. PROFESSIONAL
NEGOTIATIONS IN
EDUCATION: A
BARGAINING GAME;
1968; USA
J. Horvat; University Council for
Educational Administration, C. E. Merrill
Publishing Co
Educational administrators
1+; 2 days +

P113. PRO FOOTBALL; USA
—; Sports Illustrated Games
General
2

P114. PROGRAM DEVELOPMENT;
USA
W. H. Groff; Temple University
Post-graduate
1+
Practice in programme creation

P115. PROGRAMME PLANNING
SIMULATION; 1970; UK
R. H. R. Armstrong, M. Hobson; Birmingham
University
Post-experience, post-graduate, local
government
4+; 2 hours +
An application of NEXUS, a 'frame
technique' appropriate to many
situations, here used in programme
planning

P116. PROJECT SIERRA; 1954;
USA
—; Rand Corporation
Military
Insight into tactics of limited local warfare

P117.   PROMOTION; 1970; USA
A. Kaplan; Science Research Associates;
    Abt Associates Inc
Grade 8, high school, college
Insight into relationship between railway
    expansion, industrialization and
    urbanization in late 19th century

P118.   THE PROPAGANDA GAME;
        USA
R. Allen, L. Greene; Wff'n Proof Inc
High school and upwards
Practice in clear thinking; insight into
    techniques used to mould public opinion

P119.   PROSIM: A PRODUCTION
        MANAGEMENT SIMULATION;
        1969; USA
P. S. Greenlaw, M. P. Holtenstein; Didactic
    Systems Inc, International Textbook Co.
College
3+; 3 hours
Teaches basic concepts of production
    management

P120.   PROTOL; UK
−; Management Games Ltd
Supervisory management
1-20; up to 4 hours
Insight into product costing by use of
    break-even charts and price/demand
    schedules

P121.   PROVING THE POINT?; UK
−; Longman
Secondary school
Designed to help adolescents get on with
    people by developing their ability to see
    others points of view

P122.   PSW-1/SW-1 POLITICAL
        SIMULATION; USA
J. Parker, C. N. Smith, M. H. Whithed;
    Department of Political Science,
    Rensselaer Polytechnic Institute
High school, college
21-300; 4 hours +
Insight into effects of government on the
    individual

P123.   PSYCH-PATHS; 1970; USA
−; KMS Industries Inc
Primary school and upwards
Teaches colour and shape discrimination,
    strategic thinking

P124.   A PUBLIC OPINION GAME;
        1961; USA
W. P. Davison; Public Opinion Quarterly
College
Insight into workings of public opinion

P125.   PURCHASING; 1968; USA
E. Rausch; Didactic Systems Inc, Science
    Research Associates
Management, experienced buyers
3-5; 2 hours
Insight into major functions of purchasing

P126.   PURDUE DAIRY MANAGE-
        MENT GAME; 1966; USA
E. M. Babb, L. M. Eisgruber; Educational
    Methods Inc
College
2-16; 2 hours
Practice in business planning techniques;
    insight into the economic principles and
    characteristics of the firm and industry

P127.   PURDUE FARM MANAGE-
        MENT GAME; 1966; USA
E. M. Babb, L. M. Eisgruber; Educational
    Methods Inc
College
2 hours
Practice in decision-making in farm business
    environment

P128.   PURDUE FARM SUPPLY
        BUSINESS MANAGEMENT
        GAME; 1966; USA
E. M. Babbs, L. M. Eisgruber; Educational
    Methods Inc
College
1-16; 2 hours
Practice in business planning decisions,
    economic and accounting principles and
    characteristics of a firm or industry

P129. PURDUE FOREST
MANAGEMENT GAME;
1970; USA
B. B. Bare; Center for Quantitative Science,
University of Washington
College, graduate school
3-15; ½-2 hours
Insight into the way biological and economic
factors interact to effect the behaviour
of the forest system

P130. PURDUE SUPERMARKET
MANAGEMENT GAME;
1966; USA
E. M. Balob, L. M. Eisgruber; Educational
Methods Inc
College
2-20; 1 hour
Teaches business planning techniques,
economic and accounting principles

P131. PURSUIT; USA
–; Readers' Digest Services Inc
Junior and senior high school
Up to 30
Illustrates social change and how it occurs

P132. PUZZLE PYRAMID; 1969;
USA
E. Gilner; New Century
Primary school
1; 20-30 minutes
Practice in mastery of colour, pattern, size
and shape discrimination skills

P133. PYRAMYSTERY; Denmark
–; Piet Hein
General

Q1. QUANDARY; UK
–; Spears Games
10 years old and upwards
2

Q2. QUALITY CONTROL
MANAGEMENT; 1970; USA
–; Didactic Systems Inc
College, management
3+; 2-3½ hours
Practice decision-making in fixing priorities,
cost trade-offs, setting budgets, preparing
proceedures and setting performance
standards

Q3. QUBICK: THREE-DIMENSIONAL
TICK-TAC-TOE; USA
H. R. Leiberman; Didactic Systems Inc
High school
Insight into group planning and problem-
solving

Q4. QUEBEC 1759; USA
Gamma Two Games; Simulations
Publications
1 hour
French/British conflict in Canada

Q5. QUERIES AND THEORIES;
1970; USA
L. Allen, P. Kugel, J. Ross; Wff'n Proof Inc
Teenagers to adults
2-4
Practice in use of scientific method and
generative grammar

Q6. QUICK-SANE; USA
J. O'Neil; Wff'n Proof Inc
High school
1
Insight into topological problems

Q7. QUIET ON THE SET–TAKE
ONE; USA
P. A. Mardney; Entelek Inc
High school, college
1-6
Teaches stage terminology, types of
production problems, use of stage
equipment, personnel and locations

Q8.   QUINTO; USA
−; 3M Company
General
2-4; 40 minutes
Board game with numbered tiles; players
    compete to lay down highest score

R1.   RUSH BROOK STORYPACK;
    UK
−; Evans Ltd
9-14 years old
Teaches English through a story set in a
    small village

R2.   RUTILE AND THE BEACH;
    1970; USA
−; Macmillan Co; High School Geography
    Project
High school
27; 5-6 days
Insight into the allocation of resources and
    conflict between mining and preservation
    interests in Australia

R3.   R PRACTICE; 1962; USA
L. E. Allen; Wff'n Proof Inc
Grade 1 to college
2+
Teaches use of reiteration rule of logic

R4.   RADICALS VERSUS TORIES;
    1969; USA
D. Dal Porto; Mount Pleasant High School
High school
30-90; 4 days
Teaches that movement for US independence
    was not unanimously supported; gives
    insight into pressures involved in
    decision-making

R5.   RADIO COVINGHAM; 1972;
    UK
J. K. Jones; BBC
Secondary school
Insight into local radio and local issues

R6.   RAID; USA
C. C. Abt, R. Glazier, M. K. Moore; Games
    Central
12-16 years old
5-15
Insight into problems of and possible
    solutions to city crime racketeering;
    motivation for learning of basic maths
    and verbal skills

R7.   RAILROAD GAME; 1967; USA
F. M. Newmann, D. W. Oliver; American
    Education Publications
Junior high, high school
10-35; 1-3 hours
Insight into and sympathy for points of view
    differing from one's own

R8.   THE RAILWAY PIONEERS
    GAME; 1969; UK
R. Walford; Longman
Secondary school
Insight into the problems of railway
    builders, the role of chance in railway
    development in USA; practice in
    discussion and cooperative decision-
    making; economics and geography

R9.   RAINBOW TOWERS; UK
−; Spears Games
4-7 years old
Teaches colour discrimination and counting

R10.   REACTION; 1973; UK
−; IPC Business and Industrial Training
Supervisors
2-6
Develops and improves skills in tackling
    everyday problems

R11.   REACTION; UK
P. W. Betts; Guardian Business Services
Supervisors
6-10
Teaches problem solving

R12. READ AND SAY VERB GAME; 1957; USA
E. W. Dolch; Garrard Publishing Co
Grade 3 and upwards
2-4; 10-15 minutes
Improves reading skills and develops habit of correct word usage

R13. READING FOOTBALL; 1970; UK
Community Service Volunteers
Immigrants
Reading practice

R14. READING LABORATORY 1; USA
D. H. Parker, G. Scannele; Science Research Associates
Primary school, any pupils needing reading tuition
The 'laboratory' provides a basic grounding in the phonic structures of the language; the kit contains 44 word games (2 players each) covering 136 phonic and structural-analysis skills in 12 areas

R15. REAL NUMBERS; 1966; USA
L. E. Allen; Wff'n Proof Inc, International Learning Corp
Primary, junior high school
2+; 10-60 minutes
Teaches concepts essential to equations

R16. REAL NUMBERS (INTEGERS); 1966; USA
L. E. Allen; Wff'n Proof Inc, International Learning Corp
Primary, junior high school
2+; 10-60 minutes
Teaches concepts essential to equations

R17. REAL NUMBERS (IRRATIONAL NUMBERS); 1966; USA
L. E. Allen; Wff'n Proof Inc, International Learning Corp
Primary, junior high school
2+; 10-60 minutes
Teaches concepts essential to equations

R18. REAL NUMBERS (NATURAL NUMBERS); 1966; USA
L. E. Allen; Wff'n Proof Inc, International Learning Corp
Primary, junior high school
2+; 10-60 minutes
Teaches concepts essential to equations

R19. REAL NUMBERS (REAL NUMBERS); 1966; USA
L. E. Allen; Wff'n Proof Inc, International Learning Corporation
Primary, junior high school
2+; 10-60 minutes

R20. RECONSTRUCTION; 1970; USA
A. Kaplan et al.; Science Research Associates; Abt Associates Inc
Grade 8—high school, junior college
Insight into roles of planters, farmers and freedmen during reconstruction

R21. RECKON (POLYHEDRAL MATHEMATICS); 1970; USA
—; MASCO
High school
1-8
Develops mathematical perception, skill in basic maths processes

R22. RED DESERT; 1971; UK
J. K. Jones; J. K. Jones
Secondary school
Insight into problems of survival

R23. RED STAR—WHITE STAR; 1972; USA
—; Simulations Publications
General
2
Tactical combat in Europe in the 1970s

R24. REDI-READ 1 (THE HOME), 1970; USA
R. W. Allen; International Learning Corp
Kindergarten to grade 4
2-4; 1-10 hours
Teaches and gives practice in word recognition

R25. REDI-READ 2 (THE
       SCHOOL); 1970; USA
R. W. Allen; International Learning Corp
Kindergarten to grade 4
2-4; 2-10 hours
Teaches and gives practice in word
  recognition

R26. THE REDWOOD
       CONTROVERSY; USA
–; Houghton Mifflin Co
Junior high school
Up to 30
Insight into the conflict between ecological,
  financial and political interests during a
  Senate hearing of proposals for a
  National Park

R27. REGATTA; USA
–; 3M Co
General
Yacht racing

R28. REGULAR DIFFUSION; UK
J. P. Cole; Nottingham University
Secondary school
1-2
Insight into the spread of a new idea or
  innovation

R29. REGULAR MEETING OF
       THE WHEATVILLE BOARD
       OF EDUCATION; USA
University Council for Educational
  Administration
Educational administrators

R30. REMAGEN BRIDGEHEAD;
       1969; USA
–; Simulations Publications Inc
General
World War II

R31. REMEMBER, REMEMBER;
       1972; UK
K. Townsend; Galt Toys
Primary school
Practice in observation and memory

R32. REMINGTON RAND'S
       PURCHASING GAME; USA
–; Remington Rand
Management
Purchasing

R33. RENAISSANCE OF INFANTRY;
       1970; USA
Simulations Publications
General, military
Insight into warfare in 1250 to 1550

R34. REORGANISATION; 1970; USA
J. Zif, A. Walker, E. Orbach, H. Schwartz;
  MacMillan Co
College, management
12-21; 2½-6 hours
Insight into organizational structure,
  experience in corporate decision-making,
  relationship between strategy and structure

R35. REPRESENTATIVES AND
       ROLL CALLS: A COMPUTER
       SIMULATION OF VOTING IN
       THE HOUSE OF
       REPRESENTATIVES; 1969; USA
C. Cherryholmes, M. J. Shapiro; Bobbs-
  Merrill Inc
College
Insight into voting behaviour

R36. REPUBLICAN NATIONAL
       NOMINATING CONVENTION;
       1968; USA
–; Cedar Rapids Community School
  District
High school
150 (in 6 groups); 15 days
Insight into experiences, techniques and
  excitement of participation in a political
event

R37. RESOURCES AND ARMS;
       USA
A. Etzioni; Columbia University; Allyn
  & Bacon Inc
High school
2
Resources planning and allocation

R38. RESPONDING TO STUDENT
      STIMULI; USA
J. F. Ahern; College of Education,
  University of Toledo
College
14; 10-100 minutes
Insight into responses which may lead to
  conflict

R39. RETAIL DEPARTMENT
      MANAGEMENT GAME; USA
J. R. Green, R. L. Sisson; Didactic
  Systems Inc, J. Wiley & Son
Management
1-5
Insight into forecasting market demand,
  inventory and personnel requirements

R40. REVOLUTION; USA
−; Abt Associates Inc
Junior high school
Insight into the causes of the English Civil
  War

R41. THE RIA MANAG-
      MENT GAME; USA
−; Didactic Systems Inc.
Management
1
Insight into the management of salesmen

R42. RISK; USA
−; Palitoy
General
War game

R43. ROAD A BILITY; USA
R. Allen, J. H. Walsh; Instructional
  Learning Corp
Primary, junior high, high school
2-4; 1-8 hours
Teaches rules and regulations of driving

R44. THE ROAD GAME; 1970;
      USA
T. E. Lineham, B. E. Long; Herder and
  Herder
Primary, junior high, high school, college
8-36; ¾-2 hours
Insight into competitive behaviour

R45. ROAD GAME I; 1967; USA
H. A. Springle; Science Research Associates
Pre-school to grade 1
1-4; 20-30 minutes
Practice in spatial judgement

R46. ROAD GAME II; 1967; USA
H. A. Springle; Science Research Associates
Pre-school to grade 1
1-4; 20-30 minutes
Practice in spatial judgement

R47. ROAD PAIRS; 1972; UK
M. Lacome, C. MacLean; Galt Toys
6-12 years old
Teaches recognition of road signs

R48. THE ROAD SIGNS GAME; UK
−; Journey Games
Primary school
Recognition of road signs

R49. ROARING CAMP; USA
−; Project Simile
High school
18-35; 5-6 hours
Insight into a 19th-century mining
  community in Western USA

R50. THE ROCKY MOUNTAIN
      CURRICULUM GAME; Canada
P. S. Gillespie; University of Lethbridge
College
9-12; 1-2 hours
Insight into problems of curriculum design

R51. ROLLO BOLLO JUMBO;
      Holland
−; Jumbo Games
Primary school
Teaches number

R52. ROSE BOWL COMPUTERIZED
      FOOTBALL; USA
−; E. S. Lowe
General
2

R53. ROULETTE; Denmark
−; Piet Hein
General

R54.   ROYALTY WORD GAME;
      1964; USA
S. J. Miller; S. J. Miller Co Inc
Primary, junior high and high schools
2-4; ½-1½ hours
Spelling practice

R55.   RSVP; UK
—; Spears Games
General
2
Vocabulary and spelling exercise

R56.   RULES FOR WESTERN GUN-
      FIGHTS; 1971; UK
S. Curtis; S. Curtis
General

S1.   SAGA; 1969; UK
—; W. H. Smith
General
Insight into British history

S2.   SAGA HISTORY GAME;
      1972; UK
—; E. J. Arnold
General
2-6
Insight into important periods of history

S3.   SALE-PLAN; 1973; UK
Management Games (from an idea by
   Leyland Paint and Wallpaper Co Ltd);
   Management Games Ltd
Sales or marketing managers, salesmen
2-12; 4 hours
A competitive game on territory planning
   for salesmen

S4.   SALERNO; USA
—; Third Millenia Inc
General
World War II

S5.   SALES GAME; 1968; USA
M. Hanan; Hanan & Sons
Management
1+; 30-60 minutes
Practice in sales planning and performance

S6.   SALESPLAN; 1962; USA
M. Hanan; Hanan & Son
Management
15-45; 7 hours
Increase knowledge and practice in skills of
   sales management

S7.   SALES PROMOTION; 1968;
      USA
E. Rausch; Didactic Systems Inc
College, management, sales staff
3+; 2-3 hours
Insight into major opportunities in sales
   promotion

S8.   SALESQUOTA; USA
—; Education Research
College, management
Planning a salesman's day's work; insight
   into selling

S9.   SALES STRATEGY: 1968;
      USA
E. Rausch; Didactic Systems Inc
College, management, sales staff
3+; 1½-2 hours
Review of practice of sales skills

S10.   SALES STRATEGY AND
      MANAGEMENT; 1970; USA
J. J. Zif, E. Orbach, W. Archey; MacMillan Co
College, management
10-19; 2-4 hours
Orientation to problems of determining
   sales strategy and methods of implementing it

S11.   SAM; UK
R. Atherton; Berkshire College of Education
10 years old and over
6+
Insight into the machine operations of a
   digital computer

S12.   SAY-IT ADDITION GAME;
      1951; USA
E. W. Dolch; Garrard Publishing Co
Grade 1
2-6; 10-15 minutes
Teaches basic arithmetic combinations

S13.   SAY-IT DIVISION GAME;
        1951; USA
E. W. Dolch; Garrard Publishing Co
Grades 2 and 3
2-6; 10-15 minutes
Teaches basic number combinations

S14.   SAY-IT MULTIPLICATION
        GAME; 1951; USA
E. W. Dolch; Garrard Publishing Co
Grades 2 or 3
2-6; 10-15 minutes
Teaches basic number combinations

S15.   SAY-IT SUBTRACTION
        GAME; 1951; USA
E. W. Dolch; Garrard Publishing Co
Grades 2 or 3
2-6
Teaches basic number combinations

S16.   SCARCITY AND ALLOCATION;
        1968; USA
E. Rausch, H. J. Cranmer; Science Research
Associates; Didactic Systems Inc
High school, college
3+; 2-5 periods
Demonstrates meanings of and relationship
between saving and investment

S17.   SCAT CAT; USA
P. Lamb; Lyons & Carnahan
Grades 1-6
2-8
Beginning and ending consonants and short
vowel sounds, graphemes based on these
and graphemic applications

S18.   SCHWERPUNKT; 1969; UK
–; Strategy & Tactics
General
Insight into armoured warfare

S19.   SCIENCE MAPS; 1972; UK
G. I. Gibbs; G. I. Gibbs
6 years old and over
1+; 10 minutes +
Insight into the basic concepts of science
and their classification

S20.   SCOTICE SCRIPTI; UK
D. Miller; Albion
General
8
Irish conflict in the 11th century
(Diplomacy variant)

S21.   SCOTS; 1970; UK
A. Spackman; British European Airways
Supervisors
Simulation of control of traffic-turnround
systems for airlines

S22.   SCRABBLE; 1948; UK
–; Spears Games
General
2-4
Practice in spelling and use of vocabulary

S23.   SCRABBLE (SCHOOL
        VERSION); USA
–; Selchow & Righter Co
General
2-4
Practice in vocabulary skills

S24.   SCRABBLE FOR JUNIORS
        (SCHOOL VERSION); USA
–; Selchow & Righter Co
6-12 years old
2-4
Teaches spelling

S25.   SCRIMMAGE; USA
–; Simulations Publications Inc
General
US football

S26.   SDC SCHOOL SYSTEM
        CRISIS SIMULATION; USA
R. F. Goodman, E. Bonaich, R. J. Meeker,
D. Benor, B. Clary; –
College, management
18-36; 8 hours +
Insight into complexity of local issues,
practice in resolution of such issues
and evaluation of conflict revolution
models, policy planning.

S27. SEAL HUNTING; 1969; USA
H. Kinley; Curriculum Development
Associates, MACOS, Education
Development Center
Grade 5 and upwards
5-6; 80 minutes
Insight into strategies of successful hunting
including co-operative action

S28. SEASPEED; 1972; UK
H. R. Noon; Dunchurch Industrial Staff
College
University, management
1½ days +
A company management simulation mainly
on financial aspects and production
control; computerized

S29. SECOND GALACTIC WAR;
USA
—; Third Millennia Inc
General
Space war

S30. SECONDARY PRINCIPALSHIP
GAME I: TEACHER/
ADMINISTRATOR
CONFLICT; USA
R. Ohm; University Council for Educational
Administration
Educational Administrators

S31. SECONDARY PRINCIPAL-
SHIP GAME 2: TEACHER
CONFLICT; USA
R. Ohm; University Council for
Educational Administration
Educational Administrators

S32. SECONDARY SIMULATION;
1968; UK
P. J. Tansey; Berkshire College of
Education
Education students
1
Insight into and practice of dealing with
the problems of the 'beginning' teacher

S33. SECTION; 1969; USA
Macmillan Co; High Scool Geography
Project
High school
40; 5-8 days
Insight into allocation of state budgets and
the interaction, compromise and conflict
involved

S34. SECURITY; 1964; USA
C. E. Osgood; Institute of Communications
Research, University of Illinois
High school, college
2+; 1 hour +
Demonstrates usefulness of a strategy of
graduated and reciprocated initiatives in
tension reduction; different international strategies

S35. SELECTION OF TEACHERS:
1970; USA
D. L. Bolton; University Council for
Educational Administration
Management
1; 2 hours-1 week
Practice in position analysis, statement of
job profiles and individual profile,
predict from individual profile, select
in rank order

S36. SELECTING EFFECTIVE
PEOPLE; 1969; USA
—; Didactic Systems Inc
College, management
5+; 2-3½ hours
Practice in selection and interview
techniques; gives encouragement to
preparation of hiring procedures

S37. SELL (SALES ENVIRON-
MENT LEARNING
LABORATORY); 1969; USA
D. D. McNair, A. P. West; Simulated
Environments Inc
Management
4-100; 1-5 hours
Insight into and practice of skills in
maximizing profit, value, potential, etc
in a sales district

S38. SELLOTAPE AIRLINE
    ADVENTURE GAME; UK
−; Sellotape Ltd
Secondary school
Insight into operation of airlines using
  international air routes

S39. SENTENCE BUILDERS;
    1968; USA
L. V. Holland; Cadaco Inc
General
2-4; 30-60 minutes
Teaches sentence structure

S40. SENTENCE TRAIN; USA
P. McKee, M. L. Harrison; Houghton
  Mifflin; Abt Associates Inc
Pre-reading
2
Practice in recognition of high frequency
  words and sentence construction

S41. SENTENCING; 1972; UK
School Council; Penguin Books
Secondary school, college
Insight into treatment of crime and
  criminals

S42. SEPEX; USA
−; Central Michigan Education Research
    Council; Abt Associates Inc
Educators, management
50-60
Insight into electronics studies, feasibility
  and cost-benefit studies in large area,
  law-population districts

S43. SERMON; UK
A. Hull, B. A. Hudgell; Southgate
  Technical College
6th form and upwards
Insight into the co-relationship of economic
  factors

S44. SERPENTINO; Holland
−; Jumbo Games
Primary school
Shape matching practice

S45. SETTLE OR STRIKE; 1967;
    USA
R. Glazier, M. Rosen, J. Walker,
  I. Rosenstein; Communications
  Workers of America; Abt Associates Inc
Adults, management
7-9
Insight into collective bargaining and
  labour relations

S46. 7 UP AND DOWN; USA
P. Lamb; Lyons & Carnahan
Grades 1-6
2-7
Open-ended applications of vowel digraph
  and diphthong spelling patterns

S47. 1750 TO 1850; 1971; UK
−; Wargames Research Group
General, military
Insight into 18th and 19th-century tactics
  and warfare

S48. 1776; 1973; USA
L. Zocchi; Avalon-Hill
General
Insight into American Revolution

S49. 1787: A SIMULATION
    GAME; 1970; USA
E. Rothschild, W. Feig; Olcott Forward Inc
Junior high, high school
20; 3-10
Practice in bargaining and compromise;
  insight into the constitution and federal
  structure of the USA

S50. SEX EDUCATION; 1972; UK
−; Penguin Books
Secondary school
Insight into the provision of sex education
  in schools

S51. S.F.C.R.P. MICRO MODEL;
    1963; USA
F. Hendricks, I. Robinson, C. Gruen,
  R. Barringer, M. Ernst; A. D. Little
Trains planners to operate computer model
  of San Francisco housing market

S52. SHADY ACRES ELEMENTARY PRINCIPALSHIP; 1967; USA
J. Roberts, McIntyre; University Council for Educational Administration
Educational administrators
Practice in conflict resolution

S53. SHAKE A WFF; 1962; USA
L. E. Allen; Wff'n Proof Inc
Grade 1 to college
2-4
Practice in construction of well-formed formulae

S54. SHAKESPEARE; USA
—; Avalon-Hill Inc
8 years old and over
2-4
Learning of characters and quotations

S55. SHAKESPEARE; USA
—; Avalon-Hill Inc
General
Board game which in the advanced version depends on knowledge of Shakespeare's plays

S56. SHAPE; USA
M. S. Gordon; Abt Associates Inc
Elementary school
Practice distinguishing and constructing common geometrical forms

S57. SHAPEMASTER MINOR; UK
J. Hicks, T. Kremer; MacDonald
Primary school
2-8
Recognition of shapes, verbalization of recognition and abstraction of concepts

S58. SHAPE PUZZLE; UK
—; Spears Ltd
9 years old and over
Colour matching practice

S59. SHAPES GAME; 1972; UK
M. Ayers; Galt Toys
4½-7 years old
Shape and colour recognition practice

S60. SHIPE-SHAPE; USA
P. Lamb; Lyons and Carnahan
Grades 1-6
3-7
Teaches phoneme-grapheme relationships involving beginning or ending consonant digraphs ch, sh, th, ng

S61. SHIPS OF ALL AGES; 1972; UK
—; E. J. Arnold Ltd
Primary, secondary school
Study of ships through the ages

S62. SHIPWRECKED; 1971; UK
J. K. Jones; J. K. Jones
Secondary school
Insight into problems of survival

S63. SHOP; UK
—; Waddingtons Ltd
General
3-8
Teaches vocabulary and word and picture matching

S64. SHOP AROUND; 1972; UK
—; E. J. Arnold
Primary, secondary school
4
Practice in use of decimal currency in buying situation

S65. THE SHOPPING GAME; 1969; UK
R. Walford; Longmans Green
Secondary school
Insight into structure of shopping areas and route efficiency

S66. SHUV HA'PENNY; 1972; UK
—; Habitat
5+

S67. SIEGE; UK
—; Invicta Plastics Ltd
8 years old and over
2-4; ½ hour
Teaches simple strategic thinking

S68. SIEGE OF BODENBERG;
1967; UK
—; Strategy and Tactics
General
Insight into medieval warfare

S69. SIERRA LEONE DEVELOP-
MENTS PROJECT; 1964; USA
W. Goodman; Abt Associates Inc, West-
chester Board of Co-operative
Education Services
Grade 6
1; 4-8 hours
Knowledge of geography and history of
Sierra Leone. Insight into problems of
newly independant countries

S70. SIGMA FILE; 1973; USA
—; Seven Towns
Adult
Based on espionage, involves bluffing,
deception etc

S71. SILLY SENTENCE; USA
P. McKee, M. L. Harrison; Houghton
Mifflin; Abt Associates Inc
Pre-reading
Practice in using oral contextual clues

S72. SIM 1; USA
—; University of Southern California
Medical personnel
Training for anaesthetists

S73. SIM 2; USA
—; University of Southern California
Medical personnel
Training for anaesthetists

S74. SIMFARM; 1968; USA
W. H. Vincent; Apricultural Economics
Department, Michigan State University
College
1+; 1-7 hours
Practice in decision-making in farm
management

S75. SIMPLE DIPLOMATIC GAME;
USA
D. Benson; University of Oklahoma
High school
Insight into the relationship between action
and situation variables

S76. SIMPOL; 1970; UK
G. Mallen; System Research Ltd
Government
Insight into working of Criminal
Investigation Department

S77. SIMPOLIS; 1969; USA
C. Abt, R. Glazier et al.; Games Central
15 years old to adults
30-50
Insight into major urban problems;
transport, education, housing, civil
rights, poverty, crime and pollution

S78. SIMPORT; 1970; UK
J. Simmons; National Ports Council
Management
Port management

S79. SIMSOC; 1966; USA
W. H. Gamson; University of Michigan;
Free Press of Glencoe
High school, college, teacher trainees
20-60; 50 minutes
Insights into sociology

S80. SIMULATED COMMUNITY
TRAINING GAME; 1970; USA
F. L. Cross; R. F. Weston
College, management
1-30; 1-3 hours
Practice in decision-making

S81. SIMULATION: THE
DECISION-MAKING MODEL;
1968; USA
—; World Affiars Council of Philadelphia
High school, college
16-50; 4 hours
Insight into the elements involved in foreign
policy decision-making

S82. SIMULATION OF
AMERICAN GOVERNMENT,
1965; USA
G. M. Garvey; Kansas State Teachers
College
High school
9 + ; 2-4 hours
Insights into US political system

S83. A SIMULATION OF
COALITION PROCESSES;
1969; USA
P. M. Burgess, J. A. Robinson; Ohio State
University
Political students
Insight into roles of collective and private
benefits in voluntary associations

S84. SIMULATION OF A
COMPUTER; USA
D. E. Tritten, North Idaho Junior College
College
1-30; 1 hour +
Insight into the functioning of a computer

S85. SIMULATION OF LIFE
INSURANCE DECISIONS
(SOLID); USA
D. G. Halmstad; Metropolitan Life
Insurance Co; Society of Actuaries
Management, actuarial science trainees
5; 4 hours-3 months
Definition of technical terms and
operations; insight into importance of
goal-setting and marketing decisions

S86. SIMULATION OF SIMPLE
LEGISLATURE; 1971; Canada
C. S. Herrick; –; University of Newfoundland
High school
4-10
Insight into factors influencing a legislator's
decision-making

S87. SIMULEX; 1971; USA
D. L. Larson; University of New Hampshire
College
20-30
Integration of theory and practice of
international relations, stressing
decision-making

S88. SIMULEX: BRITISH ENTRY
INTO THE COMMON
MARKET; 1968; USA
J. L. Pyke; University of New Hampshire;
North East International Studies
Association; New Hampshire Council on
World Affairs
College
25-30

S89. SIMULEX: CRISIS IN THE
MIDDLE EAST; 1968; USA
T. L. Hopkins; University of New Hamp-
shire; North East International Studies
Association; New Hampshire Council on
World Affairs
College
21

S90. SIMULEX: INSURGENCY IN
BOLIVIA; 1968; USA
F. N. Kibler; University of New Hampshire;
North East International Studies
Association; New Hampshire Council on
World Affairs
College
32

S91. SIMULEX: INSURGENCY IN
THAILAND; 1968; USA
C. R. Cormier et al.; University of New
Hampshire; North East International
Studies Association; New Hampshire
Council on World Affairs
College
32-34

S92. SIMULEX: KASHMIR
DISPUTE; 1968; USA
R. M. Bunker; University of New Hampshire;
North East International Studies
Association; New Hampshire Council on
World Affairs
College
29

S93.   SIMULEX: PANAMA CANAL
       CONTROVERSY; 1968; USA
R. E. Molan; University of New Hampshire,
North East International Studies
Association, New Hampshire Council on
World Affairs
College
24

S94.   SIMULEX: RHODESIA,
       UNILATERAL DECLARA-
       TION OF INDEPENDENCE;
       1968; USA
M. E. Campbell; University of New
Hampshire, North East International
Studies Association, New Hampshire
Council on World Affairs
College
32-34

S95.   SIMULEX: SUCCESSION
       IN YUGOSLAVIA; 1968; USA
A. R. Cinquegrama; University of New
Hampshire, North East International
Studies Association, New Hampshire
Council on World Affairs
College
20-24

S96.   SIMULEX: VIOLATION OF
       THE TWELVE-MILE FISHERY
       LIMIT; 1968; USA
T. W. Barron; University of New Hampshire,
North East International Studies
Association, New Hampshire Council on
World Affairs
College
31

S97.   SIMULEX 2: BRITISH
       ENTRY INTO COMMON
       MARKET; 1971; USA
J. L. Pyke; University of New Hampshire,
North East International Studies
Association, New Hampshire Council on
World Affairs
College
25-30
Uses a more complex model than Simulex

S98.   SIMULEX 2: CRISIS IN THE
       MIDDLE EAST; 1971; USA
T. L. Hopkins; University of New Hampshire,
North East International Studies
Association, New Hampshire Council
on World Affairs
College
21
Uses a more complex model than Simulex

S99.   SIMULEX 2: INSURGENCY
       IN BOLIVIA; 1971; USA
F. N. Kibler; University of New Hampshire,
North East International Studies
Association, New Hampshire Council on
World Affairs
College
32
Uses a more complex model than Simulex

S100. SIMULEX 2: INSURGENCY
       IN THAILAND; 1971; USA
C. R. Cornier et al.; University of New
Hampshire, North East International
Studies Association, New Hampshire
Council on World Affairs
College
32-34
Uses a more complex model than Simulex

S101. SIMULEX 2: KASHMIR
       DISPUTE; 1971; USA
R. M. Bunker; University of New Hampshire,
North East International Studies
Association, New Hampshire Council on
World Affairs
College
29
Uses a more complex model than Simulex

S102. SIMULEX 2: RHODESIA,
       UNILATERAL DECLARATION
       OF INDEPENDENCE; 1971;
       USA
M. E. Campbell; University of New Hampshire,
North East International Studies
Association, New Hampshire Council on
World Affairs
College
32-34
Uses a more complex model than Simulex

S103. SIMULEX 2: SUCCESSION
IN YUGOSLAVIA; 1971; USA
A. R. Cinquegrama; University of New
Hampshire, North East International
Studies Association, New Hampshire
Council on World Affairs
College
20-24
Uses a more complex model than Simulex

S104. SIMULEX 2: VIOLATION
OF TWELVE-MILE FISHING
LIMIT; 1971; USA
T. W. Barron; University of New Hampshire,
North East International Studies
Association, New Hampshire Council on
World Affairs
College
31
Uses a more complex model than Simulex

S105. SINAI; 1967; USA
—; Simulations Publications Inc
General
Strategy and tactics

S106. SINDI 1; 1971; USA
G. J. Hanneman; University of
Connecticut
College
Insight into stochastic process of
innovation diffusion

S107. SIN MINUS; 1970; UK
R. H. R. Armstrong, M. Hobson;
Institute of Local Government Studies,
Birmingham University
Post-experience, post-graduate, local
government
4+; 2 hours +
An introductory game explaining the
approach of NEXUS (q.v.)

S108. SINTRACC; 1968; UK
J. Bayley, C. R. Mitchell; City University
6th form, university
15-18; 2 days
Crisis simulation of developing country with
two hostile communities; insight into
small group decision-making processes

S109. SISTER MONICA'S
IN-BASKET; 1968; USA
G. L. Immegart, D. Brent; Catholic School
Journal
Teachers
Training for principals of Catholic
elementary schools

S110. SITUATION 4; UK
—; Parker Bros
8 years old
2+
Teaches visual discrimination of shape and
colour

S111. SITTE; USA
—; Western Behavioural Sciences Institute;
Simile 2
Grade 7, graduate school
10-30; 1-4½ hours
Insight into the consequences of the actions
of interest/pressure groups in local
government

S112. SKITTLE BASEBALL; USA
—; Aurora
General
2

S113. SLAVE TRADE; USA
—; Abt Associates Inc
Junior high school
Insight into the miseries endured by slaves
in 18th century

S114. SLEUTH; USA
J. Jackson; 3M Company
General
3-7;
Deductive reasoning practice

S115. SLUDGE AND CRUD; USA
—; Instructional Simulations Inc
Grades 7-11
Practice in the identification of pollutants

S116. SMALL BUSINESS
EXECUTIVE DECISION
SIMULATION; 1967; USA
A. Dale, F. May, C. Clark, P. J. Lymberopoulos,
C. Klasson; Small Business Administration
Management
Insight into operation of small businesses

S117. SMALL BUSINESS MANAGE-
MENT EXERCISES; UK
C. Loveluck, Management Games Ltd;
Management Games Ltd
6th form and upwards
6-21; 6-8 hours
An interacting game on the operation of a
small business manufacturing and selling
one product in a competitive field.
Decisions on price, quality, advertising
expenditure, production, sales forecasting,
market research, R and D

S118. SMART (SYSTEMS
MANAGERS ADMIN-
ISTRATIVE RATING TEST);
1969; USA
A. McDonough; R. J. Irwin Inc
College, management
1
Insight into priorities and the manner in
which systems studies originate and are
processed; practice in solving management
systems projects

S119. SMOG; 1970; USA
J. Anderson, M. Trilling, R. Moody,
R. Rosen; Urban Systems Inc
14 years old to adult
2-4; 1-2½ hours
Insight into problems of pollution control

S120. SNAIL GAME; UK
−; Spears Ltd
5 years old and over
2-6

S121. SNAIL TRIAL; USA
P. Lamb; Lyons & Carnahan
Grades 1-6
3-8
Teaches beginning consonant sounds and
phonemic discrimination

S122. THE SNAKE; 1972; UK
−; Mary Glasgow Publications Ltd
11-13 years old
1+
Part of 'Bon Voyage' French language
course

S123. SNAKES AND LADDERS;
1958; UK
L. W. Downes, D. Paling; Oxford University
Press
Primary
2-4
Practice of four rules; a mathematics game

S124. SNAP; 1958; UK
L. W. Downes, D. Paling; Oxford University
Press
Primary school
2
Practice of multiplication tables

S125. SNIBBO METAL PRODUCTS;
UK
C. Loveluck; Management Games Ltd
6th form and upwards
5-30; 3 hours to 4 days
A non-interacting excercise on personnel
and production decisions in a factory
making steel tubes. Production control,
scheduling, factory layout, labour control,
purchasing, inventory control and
finance

S126. SNIP SNAP; UK
−; Rupert Hart-Davis
Pre-readers
Teaches picture and word matching

S127. SNOWPLOUGH EXERCISE;
UK
M. Hayes; M. Hayes
Local government officers
Project planning

S128. SOCCERAMA; UK
−; ASL Pastimes
General
2-6
Association football

S129. SOCCERBOSS; UK
—; Philmar; Ariel
4-8
Association football

S130. SOCIETY TODAY; USA
—; Psychology Today Games
High school, college, adults
1-8
Insight into the relationships and operations
of American Society

S131. SOLDIERS; USA
D. Isby, L. Glynn; Simulations
Publications Inc
General
Insights into small unit actions in 1914-15

S132. SOLITAIRE; Denmark
—; Piet Hein
General

S133. SOLUTION FOR ACME
METAL; 1970; USA
—; MacMillan Co High School Geography
Project
High school
7-28; 3-4 days
Practice in analysis, decision-making and
compromise

S134. SORRY!; UK
—; Waddingtons Ltd
4 years old and over
2-4
Stategic thinking

S135. SOUND BINGO; 1970; UK
—; Community Service Volunteers
Immigrants
3+
Phonic practice

S136. SOUND HOUND; USA
P. Lamb; Lyons & Carnahan
Grades 1-6
3-6
Teaches ending consonants' sounds and
phonemic discrimination

S137. SPACE: A STUDY IN
HUMAN ADAPTION; 1969;
USA
Abt Associates Inc; Edcom Systems Inc
Upper elementary school
2-32; 1-3 hours
Insight into adaption to culturally and
physically different environments

S138. SPACE RACE; 1969; USA
—; Creative Studies
Junior high, high school
2-6; 20-180 hours
Teaches and gives practice in the application
of the basic laws of physics which apply
to space travel

S139. SPACE SHOOTERS; 1950; USA
F. C. Davis; Albion
General
Entrepreneurial activity in space in 21st
century

S140. THE SPANISH AMERICAN
WAR; 1969; USA
R. Lundstedt, D. Dal Porto; R. Lundstedt
High school
20-90; 4 hours
Insight into causes of Spanish-American
War, role of pressure groups in influencing
government, operation of the committee
system in Congress

S141. SPANISH CIVIL WAR; USA
—; Simulations Publications Inc
General
Insight into Spanish Civil War

S142. SPECULATE: UK
—; Waddingtons Ltd
General
2-5
Insight into stock market operation

S143. SPEED CIRCUIT; USA
—; 3M Company
General
2-6
Motor racing

S144. SPEED COP CAREERS
        GAME; 1973; UK
B. Hobson, C. Hobson, J. Hayes; C.R.A.C.
14-15 years old
Up to 6; 1½ hours
Insight into a number of different jobs
    (qualifications needed, aptitudes, scope
    and responsibility etc)

S145. SPE-LINGO; 1964; USA
A. F. Blake; International Learning
    Corporation
Primary, junior high, high school
2-6; 15-20 minutes
Improves spelling and increases vocabulary

S146. SPELL IT; USA
P. Lamb; Lyons & Carnahan
Grades 1-6
2-8
Recognition and spelling of long and short
    vowel sounds

S147. SPELL IT; 1942; USA
Cadaco Inc
Grades 4-8
1
Teachers spelling of everyday objects

S148. SPELL MASTER; UK
–; Spears Games
4-7 years old
Spelling skills practice

S149. SPELL-O-GRAMS; USA
P. Lamb; Lyons & Carnahan
Grades 1-6
2-8
Spelling and vocabulary practice through
    word-building patterns

S150. SPENDTHRIFT; 1961; USA
R. M. Greene; Dow Jones
High school, college
1+; 2 hours +
Practice in the management of family
    finances; goal-setting and value structure
    development

S151. SPILL AND SPELL; USA
–; Parker Bros
Elementary
1+
Spelling and vocabulary

S152. SPIN-A-SOUND; USA
A. W. Heilman, R. Helmkamp, A. E. Thomas,
    C. J. Carsello; Lyons & Carnahan
Grades 1-3
2+
Practice in the association of sounds and
    symbols of consonants

S153. SPIN AND WIN; USA
P. Lamb; Lyons & Carnahan
Grades 1-6
3-6
Increases phonemic discrimination of short
    vowel sounds

S154. SPIN HARD, SPIN SOFT;
        USA
A. W. Heilman, R. Helmkamp, A. E. Thomas,
    C. J. Carsello; Lyons & Carnahan
Grades 1-3
2+
Practice in recognition of hard and soft C
    and G sounds

S155. SPINNER NUMBER GAME;
        USA
–; D. C. Heath Co
Primary school
Practice in operations with fractions

S156. SPINNING ARROW; 1958; UK
L. W. Downes, D. Paling; Oxford
    University Press
Primary school
3+
Practice in basic number bonds

S157. SPITFIRE (1939-42); USA
–; Simulations Publications Inc
General
Insights into tactical air warfare

S158.   SPLOSH; 1966; UK
I. Horton; School of Management Studies,
   Polytechnic of Central London
College, junior to senior management
20-30; 2-3 days
Setting up of a company, establishment of a
   production line, marketing planning,
   sales programming, financial structure
   of an on-going medium-size business.
   Covers all aspects of management:
   human, financial, etc. and can be played
   on a computer

S159.   SPRING GREEN MOTORWAY;
        1971; UK
−; Community Service Volunteers
Secondary school, college, adults
24; 2-3 hours
Reactions to proposed motorway route

S160.   SPRINGFIELD: FIRE ENGINE
        GAME; 1971; UK
J. P. Cole; Department of Geography,
   University of Nottingham
Secondary school
Location of fire stations in best positions in
   view of random strategy of fire occurrence
   and existence of differing fire risks

S161.   SPRODS; UK
Institute of Supervisory Management,
   Management Games; Management
   Games Ltd
6th form and upwards, supervisory and
   junior managers
Up to 24; 6-7 hours
Insight into supervisory and junior management principals, decision-making on
   purchase of raw materials, stockholding,
   machine loading, costing and pricing

S162. $QUANDER; USA
−; Avalon-Hill Inc
General
Board game in which players compete to be
   first to spend a million dollars

S163.   STAGE (SIMULATION OF
        TOTAL ATOMIC GLOBAL
        EXCHANGE); USA
−; US Defence Department
Military
5 months
Insights into the strategies, tactics and
   logistics of total atomic war

S164.   STALINGRAD; 1963; USA
−; Avalon-Hill Inc
General
2+; 3-6 hours
Illustrates Operation Barbarossa, the German
   invasion of Russia in World War II;
   practice in strategic decision-making

S165.   STANFORD BANK MANAGE-
        MENT; USA
Simulator; Stanford University
College, management

S166.   STANFORD BUSINESS
        LOGISTICS GAME; 1967; USA
K. M. Ruppenthal, D. C. Whybark,
   H. A. McKinnell; School of Business,
   Stanford University
College, management, military
2-50
Insight into problems of directing logistics
   functions of a large US manufacturing
   company

S167.   STANFORD UNIVERSITY
        MODEL ECONOMY; USA
−; Stanford University
College, management

S168.   STARPOINT TERMINAL; UK
−; Esso Petroleum Ltd
Secondary school upwards to management
Study of economics and human relations
   of productivity

S169.   STARPOWER; 1969; USA
G. Shirts; Western Behavioural Sciences
   Institute; Urbandyne
All ages
18-35; 1-2 hours
Illustrates uses and abuses of power through
   manipulation of wealth

S170. STAR RIVER PROJECT; UK
–; Esso Petroleum Ltd
Secondary school and upwards to management
Insight into environmental effects of river pollution and redundant industry in Scotland

S171. START SMART; USA
P. Lamb; Lyons & Carnahan
Grades 1-6
3-7
Practice in reinforcement of beginning consonant sounds, including blends and digraphs

S172. STATE LEGISLATOR; 1970; USA
R. A. Montgomery, S. S. Fischer, R. Hiersteiner; MacMillan Co
High school, college
18-36; 3-5 hours
Practice in negotiation, compromise and strategy formation; involvement in leglislative process; insight into conflicts surrounding regional bias and into legislative structure

S173. THE STATE SYSTEM EXERCISE; 1968; USA
W. D. Coplin; Department of Political Science, Syracuse University
College
24; 2½-4 hours
Illustrates dynamics of the international systems and factors affecting stability

S174. STEAM; USA
–; Abt Associates Inc
High school
6-15; 1-2 hours
Study of the economic considerations of application of steam in coal industry in 18th century

S175. STEPS: MODEL 1; 1959; USA
R. Boguslaw, W. Pelton; System Development Corp
Management
Insights into management for supervisors

S176. STICK-TO-IT; USA
P. Lamb; Lyons & Carnahan
Grades 1-6
3-7
Practice in spelling using long and short vowel sounds

S177. THE STICKS 'N STONES GAME (PART OF SUPERVISORY SKILLS SERIES); 1969; USA
–; Training Development Center
Management
3-16; 1-2 hours
Demonstrates the personality attributes important to effective leadership; illustrates purpose and function of supervision; demonstrates difference between human and task handling skills

S178. STOCKBROKER; 1973; UK
–; Intellect (UK) Ltd
2-6
Stock market investment game

S179. STOCK EXCHANGE FINANCE GAME; UK
–; Stock Exchange
Secondary school

S180. STOCKMARKET; USA
J. M. Rowland; J. M. Rowland
Junior high, high school
3-10; 1 hour +
Insight into fundamental principles of stock market

S181. STOCK MARKET GAME; 1970; USA
Avalon-Hill Co; Avalon-Hill Co
General
2+; 45-120 minutes
Insight into operation of stock market and the rise and fall of prices

S182. THE STOCK MARKET GAME; USA
–; Didactic Systems Inc
High school, college
2-5
Insight into stock market dealings

S183. STOCKS AND BONDS; USA
—; 3M Company
10 years old and over
2-8; 1 hour
Insight into stock market operation

S184. STOP DOT; USA
P. McKee, M. L. Harrison; Houghton
   Mifflin; Abt Associates Inc
Pre-reading
2-4
Insight into left-to-right sequence, practice
   in using comma and full point

S185. STRANGE BEDFELLOWS;
   1971; USA
—; Wff'n Proof Inc
High school, college
Practice in matching quotations and authors
   in social studies

S186. STRATEGOS; Holland
—; Jumbo
General
Napoleonic period war game

S187. STRATEGOS; USA
C. Totten; Simulations Publications
General
War game

S188. STRATEGY 1; USA
—; Simulations Publications Inc
General
General insight into strategy

S189. STRATHCLYDE EXERCISE;
   1967; UK
S. Abrines; British Broadcasting Corporation
Management
2 days
Insight into industrial relations

S190. STRAT-O-MATIC BASEBALL;
   USA
—; Strat-o-Matic Games
General
2

S191. STRAT-O-MATIC FOOTBALL;
   USA
—; Strat-O-Matic Games
General
2

S192. STREET COMBAT; USA
—; Simulations Publications Inc
General
World War II game

S193. STREETS AHEAD; 1971; UK
Liverpool EPA Project; Liverpool
   Educational Priority Area Project
Secondary school, further education
Increases urban social awareness

S194. STRIKE; 1970; USA
S. Fischer, R. Montgomery, M. Kaunfer;
   Macmillan Co
High school, college
15-42; 2½ hours
Insight into complexities of labour/
   management relations, public effects of
   private enterprise, achievement of
   objectives, conflict resolution

S195. STRIKE; 1971; UK
D. Bruce; Brooklands County Technical
   College
Craft students
Insight into the complexity of the interests,
   attitudes and actions which make up an
   industrial dispute

S196. STRIKE; USA
—; Interact
7th grade to college
Up to 35
Insight into the development of labour
   relationships during the 19th century

S197. STRUCTURAL LINGUISTICS;
   USA
—; Nova High School
High school
English language practice

S198. STURMOVIK (1941-45); USA
—; Simulations Publications Inc
General
Insight into tactics of air war

S199. SUM-IT; 1968; UK
N. P. Vine; John Waddington Ltd
Primary, secondary school
2-4
Practice in adding and subtracting sums of
decimal money

S200. SUMMIT; USA
M. Bradley; Milton Bradley
General
3-6
Political strategy game

S201. SUM TIMES; USA
—; 3M Company
8-12 years old
2-4
Practice in basic maths skills

S202. SUM UP; USA
—; 3M Company
General
2-4
Practice in equations (addition, subtraction,
multiplication and division)

S203. SUMERIAN GAME; 1964; USA
M. Addis, B. Moncrieff; Westchester Board
of Cooperative Educational Services
Grade 6
1; 4-10 hours
Insight into and practice of decision-
making in an agricultural economy

S204. SUNSHINE; USA
D. Yount, P. DeKoch; Interact
Grade 5 to college
20-36; 16-22 lessons
Insight into problems and relationships of
urban areas with mixed cultural and
racial groups

S205. SUPERB; USA
J. C. Hodder, W. R. Folks; Science Research
Associates; Abt Associates Inc
Super-market executives
Practice in decision-making in supermarket
management

S206. SUPERGAME (SUPERVISORY
TRAINING GAME); USA
D. F. Williams; Bank of California
Management
5-9; 12 hours
Acquaintance of employees with the
problems of supervision at low levels

S207. SUPERMARKET; 1969; USA
—; Creative Studies Inc
Junior high school
8-40; 1-4 hours
Practice of quantitative skills in buying and
selling; multiplication of one and two
digit numbers; finding and comparing
percentages

S208. SUPERMARKET STRATEGY;
1970; USA
J. Zif, I. Ayal, E. Orbach; Creative Studies Inc;
MacMillan Co
College, graduate school, management
4-24; 6 hours +
Insight into major marketing concepts;
practice in data collection and analysis
and decision-making

S209. SUPERMAYOR; USA
—; Instructional Systems Inc
High school
Insight into city government

S210. SUPER PUZZLE: UK
—; Spears Games Ltd
7 years old to adult
2
Practice in manipulation of geometric
shapes

S211. SUPERVISION; USA
—; Education Research
College, management
Practice in supervisory skills

S212. SUPERVISORY MANAGE-
MENT EXERCISE; 1965; USA
J. N. Fairhead, D. S. Pugh, W. J. Williams;
English University Press
Management
Insight into the co-ordination and control
of a manufacturing department

S213. SUPERVISORY SKILLS;
1968; USA
E. Rausch; Didactic Systems Inc; Science
Research Associates
Junior management
3+; 2 hours
Practice of management skills, insight into
supervisory functions

S214. SUPPLEMENTAL FUNDING;
1969; USA
R. E. Horn, J. A. Sarthory, D. E. Wade;
Information Resources Inc
Management
4+; 12 hours
Teaches sources, processes, costs, risks and
potential of supplemental funding and
gives insight into complexity of relationships involved

S215. SUPRA; USA
I. T. and W. R. Folks; Science Research
Associates; Abt Associates Inc
Sales trainees
6-36
Insight into sales and buying problems

S216. SURVIVAL; 1972; UK
P. Newmark; Galt Toys
8 years old and over
Demonstrates animal life chains of
dependancy and effects of pollution

S217. SURVIVAL DIARY; 1971;
Canada
J. L. Heap; University of British Columbia
High school
Insight into reactions to escape from an
ecological apocalypse

S218. THE SURVIVAL GAME;
1971; UK
J. T. Wilson; Visual Education
College
Insights into conflicts in ethics, morals and
political expression

S219. SYLLABLE COUNT; USA
A. W. Heilman, R. Helmkamp, A. E. Thomas,
C. J. Carsello; Lyons & Carnahan
Grades 1-3
2+
Practice in hearing and recognition of
syllables and in applying principles of
syllabication

S220. SYLLABLE GAME (SLIGHT
SYLLABLE SOLITAIRE);
1948; USA
E. W. Dolch; Garrard Publishing Co
Grade 3 and upwards
1-2; 10-15 minutes
Teaches word attack, sight recognition of
syllables, rules for forming syllables;
creates habit of looking for parts of long
words; recognition of many common
words

S221. SYMMETRIC TAC-TICKLE;
1967; USA
H. D. Ruderman; Wff'n Proof Inc
High school, college
2
Practice in strategic thinking and insight
into concept of symmetry

S222. SYSTEM 1; 1969; USA
Instructional Systems Inc; Instructional
Simulations Inc
Primary, junior high, high school
2-8; 20-40 minutes
Illustrates simple and complex relationships within subject matter in a
systematic manner; can be used in any
subject

S223. SYSTEMS ENGINEERING OF
EDUCATION VII: GENERAL
SYSTEM MODEL FOR
EFFECTIVE CURRICULUMS;
1972; USA
L. C. Silvern; Education & Training
Consultants Co
Teachers
Insight into uses of models in educational
planning

T1. TABLE CRICKET: CLUB, UK
–; Subbuteo Ltd
General

T2. TABLE CRICKET: DISPLAY;
UK
–; Subbuteo Ltd
General

T3. TABLE CRICKET: TEST
MATCH; UK
–; Subbuteo Ltd
General

T4. TABLE RUGBY: INTER-
NATIONAL; UK
–; Subbuteo Ltd
General

T5. TABLE SOCCER;
CONTINENTAL CLUB; 1947;
UK
P. Adolph; Subbuteo Ltd
General

T6. TABLE SOCCER:
CONTINENTAL DISPLAY;
1947; UK
P. Adolph; Subbuteo Ltd
General

T7. TABLE SOCCER; UK
–; Waddingtons Ltd
8 years old and over
2

T8. TAC 3; USA
–; Simulations Publications Inc
General
Tactical game; forerunner to Panzerblitz

T9. TAC 13; USA
–; Strategy and Tactics
General
Strategic and tactical thinking

T10. TAC 14; USA
–; Strategy and Tactics
General
Strategic and tactical thinking

T11. TACSPIEL; USA
–; Research & Analysis Corp
Military
8 hours
Insight into land war tactics

T12. TAC-TICKLE; 1965; USA
H. D. Ruderman
–; Wff'n Proof Inc
General
2
Practice in strategic thinking

T13. TAF 4-2-4; 1972; UK
–; TAF Sports Games Ltd
General
Table football

T14. TAKE; USA
J. S. Coleman; Academic Games
Associates Inc
General
2
Teaches strategy involved in numbers

T15. TAKE (A SOUND
MATCHING GAME); 1953; USA
E. W. Dolch; Garrard Publishing Co
Grade 2 and upwards
2-4; 10-15 minutes
Teaches principles of sound through
constant repetition

T16. TAKE THE BRAIN; UK
–; Parker Bros
8 years old and over
Stategic and tactical insights

T17.   TAKTISKA RELIEF-
KRIGSSPEL; 1886; Sweden
de Ridderstad; Imprimerie Centrale
Military
Practice in matching tactics to terrain

T18.   TANNENBURG; UK
−; Simulations Publications Inc
General
Insight into World War I battle between
   Austrians, Germans and Russians

T19.   TEACH KEY MATH; USA
−; 3M Company
5-8 years old
Up to 7
Practice in solution of addition and
   subtraction problems

T20.   TEACH KEY READING
   AND SPELLING; USA
−; 3M Company
5-8 years old
Teaches spelling

T21.   THE TEACHER; 1968; UK
P. J. Tansey; Berkshire College of
   Education
Education students
Practice in problem-solving in a school
   situation

T22.   THE TEACHER SELECTION
   SIMULATION; USA
B. Bolton; University Council for
   Educational Administrators
Educational administrators

T23.   TEACHER STUDENT
   INTERACTION GAME 1;
   1970; USA
H. H. Mette; Randa Inc
College, in-service teachers
3-6; 10-60 minutes
Practice in student/teacher interaction;
   evaluation of teaching methods and
   theory

T24.   TEACHING PROBLEMS
   LABORATORY: BEING
   IMPATIENT WITH STUDENTS;
   1967; USA
D. R. Cruickshank *et al.*; Science Research
   Associates
Teachers

T25.   TEACHING PROBLEMS
   LABORATORY: BEING
   UNHAPPY WITH CLASS-
   ROOM CLERICAL WORK;
   1967; USA
D. R. Cruickshank *et al.*; Science Research
   Associates
Teachers

T26.   TEACHING PROBLEMS
   LABORATORY: CONTACTING
   AN UNRESPONSIVE PARENT;
   1967; USA
D. R. Cruickshank *et al.*; Science Research
   Associates
Teachers

T27.   TEACHING PROBLEMS
   LABORATORY:
   DIFFERENTIATING
   INSTRUCTION FOR SLOW,
   GIFTED AND AVERAGE
   CHILDREN IN THE CLASS;
   1967; USA
D. R. Cruickshank *et al.*; Science Research
   Associates
Teachers

T28.   TEACHING PROBLEMS
   LABORATORY: DISCUSSING
   WITH PARENTS THEIR
   CHILDREN'S UNSATIS-
   FACTORY ACHIEVEMENT;
   1967; USA
D. R. Cruickshank *et al.*; Science Research
   Associates
Teachers

T29. TEACHING PROBLEMS
LABORATORY: EVALUAT-
ING TEACHING
OBJECTIVES; 1967; USA
D. R. Cruickshank *et al.*; Science Research
Associates
Teachers

T30. TEACHING PROBLEMS
LABORATORY: FEELING
NERVOUS WHEN SUPER-
VISED; 1967; USA
D. R. Cruickshank *et al.*; Science Research
Associates
Teachers

T31. TEACHING PROBLEMS
LABORATORY: FEELING
UNCOMFORTABLE ABOUT
GIVING FAILING GRADES;
1967; USA
D. R. Cruickshank *et al.*; Science Research
Associates
Teachers

T32. TEACHING PROBLEMS
LABORATORY: FINDING
APPROPRIATE MATERIALS
FOR CHILDREN READING
ONE OR MORE YEARS
BELOW GRADE LEVEL;
1967; USA
D. R. Cruickshank *et al.*; Science Research
Associates
Teachers

T33. TEACHING PROBLEMS
LABORATORY: FINDING
FILMS AND FILMSTRIPS
RELATED TO AN AREA
OF STUDY; 1967; USA
D. R. Cruickshank *et al.*; Science Research
Associates
Teachers

T34. TEACHING PROBLEMS
LABORATORY: GETTING
PARENTS TO TAKE AN
INTEREST IN THEIR
CHILDREN'S SCHOOL AND
CLASS-WORK; 1967; USA
D. R. Cruickshank *et al.*; Science Research
Associates
Teachers

T35. TEACHING PROBLEMS
LABORATORY: GETTING
STUDENTS TO DO
HOMEWORK; 1967; USA
D. R. Cruickshank *et al.*; Science Research
Associates
Teachers

T36. TEACHING PROBLEMS
LABORATORY: HANDLING
CHILDREN'S AGGRESSIVE
BEHAVIOR TOWARDS
ONE ANOTHER; 1967; USA
D. R. Cruickshank *et al.*; Science Research
Associates
Teachers

T37. TEACHING PROBLEMS
LABORATORY: HANDLING
THE CONSTANTLY
DISRUPTIVE CHILD; 1967;
USA
D. R. Cruickshank *et al.*; Science
Research Associates
Teachers

T38. TEACHING PROBLEMS
LABORATORY: HAVING
A DISTASTE FOR GRADING
PAPERS; 1967; USA
D. R. Cruickshank *et al.*; Science Research
Associates
Teachers

T39. TEACHING PROBLEMS
LABORATORY: HAVING
CHILDREN DO INDEPENDENT
WORK QUIETLY; 1967; USA
D. R. Cruickshank *et al.*; Science Research
Associates
Teachers

T40. TEACHING PROBLEMS LABORATORY: HAVING STUDENTS SEE RELATIONSHIP BETWEEN UNDESIRABLE BEHAVIOR AND ITS CONSEQUENCES; 1967; USA
D. R. Cruickshank *et al.*; Science Research Associates
Teachers

T41. TEACHING PROBLEMS LABORATORY: HELPING A STUDENT WITH A DESTRUCTIVE HOME SITUATION; 1967; USA
D. R. Cruickshank *et al.*; Science Research Associates
Teachers

T42. TEACHING PROBLEMS LABORATORY: INTEGRATING THE ISOLATED, DISLIKED CHILD; 1967; USA
D. R. Cruickshank *et al.*; Science Research Associates
Teachers

T43. TEACHING PROBLEMS LABORATORY: INTERPRETING CHILDREN'S TRUE CAPABILITIES TO PARENTS; 1967; USA
D. R. Cruickshank *et al.*; Science Research Associates
Teachers

T44. TEACHING PROBLEMS LABORATORY: INVOLVING CHILDREN IN GROUP DISCUSSIONS; 1967; USA
D. R. Cruickshank *et al.*; Science Research Associates
Teachers

T45. TEACHING PROBLEMS LABORATORY: INVOLVING PUPILS IN SELF-EVALUATION; 1967; USA
D. R. Cruickshank *et al.*; Science Research Associates
Teachers

T46. TEACHING PROBLEMS LABORATORY: NOT KNOWING HOW TO DEAL WITH CHILDREN'S READING PROBLEMS; 1967; USA
D. R. Cruickshank *et al.*; Science Research Associates
Teachers

T47. TEACHING PROBLEMS LABORATORY: NOT KNOWING WHAT TO DO WITH STUDENTS WHO FINISH WORK EARLY; 1967; USA
D. R. Cruickshank *et al.*; Science Research Associates
Teachers

T48. TEACHING PROBLEMS LABORATORY: LACKING ENTHUSIASM FOR A SUBJECT; 1967; USA
D. R. Cruickshank *et al.*; Science Research Associates
Teachers

T49. TEACHING PROBLEMS LABORATORY: MOTIVATING STUDENTS TO WORK ON CLASS ASSIGNMENTS; 1967; USA
D. R. Cruickshank *et al.*; Science Research Associates
Teachers

T50. TEACHING PROBLEMS LABORATORY: PREPARING VALID CLASSROOM TESTS; 1967; USA
D. R. Cruickshank *et al.*; Science Research Associates
Teachers

T51. TEACHING PROBLEMS LABORATORY: PROVIDING APPROPRIATE WORK FOR REST OF CLASS WHILE WORKING WITH A SMALL GROUP OR WITH INDIVID- UAL CHILDREN; 1967; USA
D. R. Cruickshank *et al.*; Science Research Associates
Teachers

T52. TEACHING PROBLEMS LABORATORY: RELATING A COMPLEX SUBJECT TO CHILDREN MEANINGFULLY; 1967; USA
D. R. Cruickshank *et al.*; Science Research Associates
Teachers

T53. TEACHING PROBLEMS LABORATORY: SECURING HELP IN SELECTING INSTRUCTIONAL MATERIALS; 1967; USA
D. R. Cruickshank *et al.*; Science Research Associates
Teachers

T54. TEACHING PROBLEMS LABORATORY: TELLING PARENTS THAT THEIR CHILDREN HAVE SERIOUS PROBLEMS; 1967; USA
D. R. Cruickshank *et al.*; Science Research Associates
Teachers

T55. TEAMSKILL; 1973; UK
—; Honeywell Ltd; Institute of Production Engineers
College, management
1-48
Production process management

T56. TELECITY; USA
—; Applied Simulations International Inc
High school
25-75
To increase level of awareness so that a holistic outlook on problems is achieved

T57. TEMPER (TECHNOLOGICAL, ECONOMIC, MILITARY, POLITICAL EVALUATION ROUTINE); 1966; USA
C. Abt; University of Pennsylvania; Raytheon; Industrial College of the Armed Forces
Military
General simulation structure designed to handle different theories of international relations in cold war and give insights and a result

T58. TENEMENT; 1972; UK
G. Cooper; Shelter
14 years old and over
14-30; 1½-2 hours
Insight into the problems of families living in a multi-occupied house

T59. THE TEN GAME; 1957; USA
E. W. Dolch; Garrard Publishing Co
Grades 1-2
2-6; 10-15 minutes
Practice in the meaning and construction of base ten numbers

T60. TERRITORIES GAME; 1965; USA
G. H. Shure, R. J. Meeker; System Development Corp
College
2
Practice in minimizing losses and costs while maximizing possession of territory

T61. TETRAD CARD GAME; 1972; UK
—; E. J. Arnold Ltd
Primary school
2-5
Teaches geometrical figures

T62. THEATERSPIEL; USA
—; Research & Analysis Corp
Military
Insight into war theatre logistic problems

T63.   THEATRE TICKETS; 1967;
       USA
H. A. Springle; Science Research Associates
Pre-school to grade 1
1-4; 20-30 minutes
Insight into concepts of 'more than' and
   'less than'

T64.   THEY SHOOT MARBLES,
       DON'T THEY?; USA
F. Goodman; University of Michigan
College
Insight into the problems and dynamics of
   community relations

T65.   THINKING MAN'S FOOTBALL;
       USA
—; 3M Company
General
2
American football

T66.   THINKING MAN'S GOLF;
       USA
—; 3M Company
General
1-4

T67.   THIRD AGE; UK
J. Piggott; J. Piggott
General
Diplomacy variant (Tolkien)

T68.   THE THIRD AIRPORT;
       1971; UK
—; British Broadcasting Corporation
General
Simulates inquiry into siting of a third
   London airport

T69.   THE THREE GAME; 1967;
       USA
H. A. Springle; Science Research Associates
Pre-school to grade 1
1-4; 20-30 minutes
Insight into nature of equivalence

T70.   3-DOX
—; Heals Ltd
General
Three dimensional 'noughts and crosses'

T71.   30 YEARS WAR; USA
—; Simulations Publications Inc
General

T72.   T.I. BUSINESS GAME;
       1969; UK
Tube Investments Personnel Department;
   Tube Investments
College, management trainees
15 upwards; 1 day
A production management game covering
   ordering, stock control, product selection,
   machine loading, scheduling etc. also
   general business decision-making

T73.   TIMSIM: A COMPUTERISED
       MANAGEMENT GAME FOR
       TRAVEL INDUSTRY
       MANAGEMENT; 1969; USA
L. Jacobs; L. Jacobs
College, graduate school, management
Up to 100; 3 months
Practice in the use of information in
   decision-making

T74.   TOLL ROAD; USA
P. Lamb; Lyons & Carnahan
Grades 1-6
3-8
Reinforces morphemic principles

T75.   TOOL ROOM GAME;
       1960; USA
F. M. Campbell, E. R. Attworth; Booz,
   Allen and Hamilton
College, management
1+; 50-60 minutes
Demonstrates interacting nature of costs
   arising from tool-room operating
   decisions

T76.   TOOLS AND TECHNOLOGY;
       USA
—; Edcom Systems Inc, Abt Associates Inc
Grade 4
Illustrates, by problem-solving, the
   relationship of Eskimo culture to an
   Arctic environment

T77.   THE TOP MAN GAME; USA
A. N. Schrieber; Graduate School of
  Business Administration, University of
  Washington
College, graduate school, management
6-36; 8-40 class periods
Insights into and practice of skills of
  dynamic decision-making

T78.   TOP MANAGEMENT
       EXERCISE; 1965; UK
J. McGregor, J. W. Fairhead, D. S. Pugh,
  W. J. Williams; English Universities Press
Management
Experience of working under differing
  personalities; insight into discrepancy
  between responsibility and authority.
  Experience of operating at policy-making
  level; insight into significant variables
  and their relationships to each area of
  business

T79.   TOP MANAGEMENT
       SIMULATION; 1970; USA
R. M. Hodgetts; University of Nebraska
College, graduate school, management
24; 12-20 hours
Insight into interaction between teams and
  functions

T80.   TOP OPERATING MANAGE-
       MENT GAME; USA
J. R. Greene, R. L. Sisson; Didactic
  Systems Inc; J. Wiley & Sons
Management
3+
Decision-making practice

T81.   TORESIDE COMPREHENSIVE
       SCHOOL; UK
S. Ferguson; School of Education, Liver-
  pool University
Education students, teachers
Insight into relationships between teachers
  in a school staff room

T82.   TORLONIA IRON AND
       STEEL WORKS; 1971; UK
J. P. Cole; Department of Geography,
  Nottingham University
Secondary school upwards
Location of a new iron and steel works,
  taking account of market needs, resources
  and transportation available

T83.   TOTOPOLY; UK
–; Waddingtons Ltd
8 years old and over
2-6
Horse racing

T84.   TOT-TEN; UK
–; Spears Games
6 years old up to adult
2-6
Teaches number bonds and gives practice in
  addition

T85.   TOWER GAME; Denmark
–; Piet Hein
General

T86.   TOWER OF BRAHMA; UK
–; Waddingtons Ltd
8 years old and over
1
Teaches strategic thinking

T87.   TRACKING; USA
–; Edcom Systems Inc; Abt Associates Inc
Grade 4
Insight into how Aranda survive the harsh
  Australian outback conditions with a
  limited technology

T88.   TRACTS; 1969; USA
–; Instructional Simulations Inc
Junior high, high school, college,
  management
12-20; 2-4 hours
Practice in strategic and logistic thinking in
  urban land development; insight into
  problems of intra-city land usage

T89.   TRADE AND DEVELOP;
1969; USA
S. Livingston; Academic Games Associates Inc;
Didactic Systems Inc
Upper elementary, high school
4-10
Insight into processes of economic development and understanding of basic
economic concepts

T90.   TRADE THE MARKET, USA
—; Didactic Systems Inc
High school, college
2+
Insight into the stock market

T91.   THE TRADE TRIANGLE
GAME; 1971; UK
M. Renwick, H. Tolley; Jackdaw
Publications Ltd
Primary upwards
2+
Insight into operation of 18th century
North Atlantic trade triangle

T92.   TRAFALGAR; 1969; USA
—; Simulations Publications Inc
General
Naval strategy and tactics

T93.   TRAINING OFFICER;
1968; UK
K. Jackson; Spencer & Halstead Ltd
Trainee training officers
1
Insight, through his correspondence, into
the problems which face a training
officer

T94.   TRAINING OF TEACHER
GROUP LEADERS; 1972; UK
C. Longley; British Broadcasting Corporation
Teachers
6; 1 hour
Insight into ROSLA (raising of school
leaving age)

T95.   TRANSACTION; USA
—; Didactic Systems Inc
High school, college
Insight into the operation of the stock
market

T96.   TRANSACTION: THE
AUTHENTIC STOCK
MARKET GAME; 1962, 1968;
USA
J. R. Tusson; Study-Craft Educational
Products, Entelek Inc
Junior high, high school, college, graduate
school
2-25; 1 hour +
Insight into and practice in stock market
trading

T97.   TRANSPORTATION 1; 1969;
USA
H. A. Springle; Science Research Associates
Pre-school to grade 1
1-4; 20-30 minutes
Transport

T98.   TRANSPORTATION 2;
1969; USA
H. A. Springle; Science Research Associates
Pre-school to grade 1
1-4; 20-30 minutes
Transport

T99.   TRANSPORTATION; USA
S. Bornstein; Abt Associates Inc
High school
Transport

T100.   TRAPPER; UK
—; Mosesson Games Ltd
Adults
3-9
Gives practice in elementary mathematics

T101.   TREK; UK
—; Spears Games
10 years old and over
3-6
Insight into problems of transportation in
undeveloped areas

T102. TRIANGLE TRADE; 1969;
    USA
R. Durham, J. Crawford, P. Twelker;
    Teaching Research Division, Oregon
    State System of Higher Education
High school
15-44; 50-150 minutes
Shows how New England did not fit into
    the British mercantile system and why
    British government opposed 'trade
    triangle'. Relationship of British
    mercantile system to economic develop-
    ment of New England

T103. TRIG; UK
Management Games; Management Games Ltd
6th form, college, university, management
1-21; 3-5 hours
A competitive, non-interacting game on
    production management; decisions on
    materials purchasing, machine allocation,
    stock control, costing and pricing

T104. TRI-NIM; USA
H. Hicks, B. Hicks; Wff'n Proof Inc
General
A version of 'Nim' (the 'Marienbad game')

T105. A TRIP TO THE MOON;
    1968; USA
L. G. Gotkin, E. Richardson; New Century
Kindergarten
4+; 15 minutes
Teaches reading readiness skills; promotes
    comprehension, both visual and verbal,
    and correct use of adjectives

T106. TRIPPPLES; USA
—; Benassi Enterprises
General
Feedback strategy game

T107. TRIREME; UK
Decalset
—; General
Insight into naval tactics in ancient Greece

T108. TROMINO-GO; 1969; USA
S. Sackson; S. Sackson
Version of dominoes

T109. TUF; USA
—; Avalon-Hill Inc
8 years old and over
5
Practice in equation forming

T110. TUFABET; 1969; USA
J. Brett, P. Brett; Avalon-Hill Inc
General (8 years old and over)
1-4; 2-10 minutes
Gives vocabulary and spelling practice

T111. TUFTY ROAD SAFETY
    GAME; 1972; UK
—; E. J. Arnold Ltd
Primary school
Teaches road safety

T112. TWIN-EARTHS DIPLOMACY;
    USA
D. Miller; Albion
General
Diplomacy variant

T113. TWITCHIT; 1972; Denmark
Piet Hein; Piet Hein
General
1
Dodecahedron with rotating faces; aim is to
    turn them until three different symbols are
    together at each corner

T114. 12 O'CLOCK HIGH; USA
—; Simulations Publications Inc
General

T115. TWIXT; USA
—; 3M Company
General
2 (or 2 teams); 30 minutes
Tactical board game

T116. THE TWO GAME; 1967; USA
H. A. Springle; Science Research Associates
Pre-school to grade 1
1-4; 20-30 minutes
Insight into nature of equivalences

U1.   U-BOAT; USA
—; Avalon-Hill Inc
General
Insight into submarine war tactics

U2.   THE UCLA EXECUTIVE
DECISION GAME; USA
—; U.C.L.A.
College, management

U3.   ULCERS; 1973; UK
—; Waddingtons Ltd
9 years old and over
2-4
Hiring, promoting and 'poaching' of office
staff to build up a full-strength company

U4.   ULSTER IN YOUR HANDS;
UK
J. McCormick; Schools Council
Secondary school
Insight into complexities of Government
decision-making in Northern Ireland

U5.   ULSTER POLITICS; 1969; UK
D. J. Unwin, J. McCormick; Portadown
College, University of Ulster
6th Form
Insight into the problems of Northern
Ireland

U6.   UNC RETAIL GAME; 1965;
USA
R. E. Schellenberger; Graduate School of
Business, University of North Carolina
College, graduate school, vocational
3-9; 4-20 hours
Practice in dealing with management
problems, roles and systematic techniques

U7.   UNITEX; 1967; USA
H. E. Johnson; College of Business
Administration, University of Texas
College
1+; Several days
Practice skills of production and inventory
control and use of computer simulation
for problem analysis

U8.   UNIVERSITY FACULTY
RECRUITMENT; 1971; USA
J. B. Gunnell; Ohio State University
University Administrators
Insight into aspects of recruitment
strategy

U9.   UNIVERSITY OF CHICAGO
STOCK MARKET GAME;
USA
—; University of Chicago
High school, college

U10.   UNIVERSITY OF NEVADA
MANAGEMENT GAME;
USA
—; University of Nevada
College, management

U11.   UNIVERSITY OF TEXAS
METAL-WORKING
SIMULATION; USA
—; University of Texas
College, management

U12.   UPPOE (UNIVERSITY OF
PITTSBURGH PRODUCTION
ORGANISATION EXERCISE);
1965; USA
B. Bass, J. A. Vaughan; Instad Ltd
Graduate school, management
16-128; 1-5 days
Uses players' experience to focus a wide
variety of behavioural phenomena
important in organizing and managing

U13.   URBAN DYNAMICS; 1971;
USA
L. A. Callahan et al.; Urbandyne
High school, college, adults
12-20
Insight into urban structures, problems and
possibilities; shows inadequacy of
simplistic solutions

U14.   U.S.N.; USA
J. Dunnigan; Simulations Publications Inc
General
Insight into strategy in Pacific naval war-
fare 1941-1943

V1. THE VALUE GAME; 1970; USA
T. E. Lineham, W. S. Irving; Herder & Herder
High school, college
5-35; ¾-1½ hours
Demonstrates lack of ethical consensus in random group and the inadequacy of a system advocating absolute right and wrong

V2. VECTOR; USA
—; Plan B Corporation
General
Tactical board game

V3. VENTURE; USA
S. Sackson; 3M Company
General
2-6
Insight into the formation of business conglomerates

V4. VENTURE; 1966; USA
Procter and Gamble Co; Procter and Gamble Co
High school, college
20
Insight into business operations in a competitive economy

V5. THE VERBAL GAME; 1968; USA
R. Brown; MACSCO
Junior high, high school
1-8
Practice in vocabulary, word building and sentence sense

V6. VERDICT II; USA
—; Avalon-Hill Inc
General
Trial procedure

V7. VERDUN; USA
—; Simulations Publications Inc
General
World War I

V8. VIETNAM; 1967; USA
—; Simulations Publications Inc
General
Vietnam war

V9. VIETNAM CRISIS SIMULATION; UK
—; Bingley College of Education
Students

V10. VIETNAM 72; 1972; USA
—; Simulations Publications Inc
General
Vietnam war

V11. THE VIETNAM WAR GAME; 1966; UK
J. Laulicht, J. Martin; New Society
General
Vietnam

V12. VISA; UK
—; Urwick Orr
Supervisors
Practice in supervisory control

V13. VOTES; USA
—; El Capitan High School
High school
Political insights

V14. VOWEL DOMINOES; USA
A. W. Heilman, R. Helmkamp, A. E. Thomas, C. J. Corsello; Lyons and Carnahan
Grades 1-3
2+
Practice auditory discrimination of long and short vowels using diacritical marks

V15. VOWEL LOTTO; 1956; USA
E. W. Dolch; Garrard Publishing Co
Grade 1 upwards
2-10; 10-15 minutes
Practice vowel listening and use

W1. WALT DISNEY DOMINOES; UK
—; Waddingtons Ltd
3 years old and over
2-6
Teaches number and four rules

W2. WANTED; 1972; UK
−; Spears Games
8 years old and over
2-4
Memory training

W3. WAR IN THE EAST; USA
−; Simulations Publications Inc
General
Insight into Russian campaign in World
    War II

W4. WARRI; Denmark
−; Piet Hein
General

W5. WARLORD; 1972; UK
M. Hayes; M. Hayes
General
World strategy war game

W6. WAR OR PEACE; 1966; USA
J. D. Gearon; Social Education
15 years old and upwards
7+
Insight into international crises and political
    concepts such as 'balance of power'

W7. WARS OF NAPOLEON; USA
−; Simulations Publications Inc
General

W8. WASHINGTON UNIVERSITY
    BUSINESS GAME; 1969; USA
P. Niland, J. W. Towle, C. A. Danten;
    Graduate School of Business Administra-
    tion, Washington University
Graduate school, management
12-72; 4-10 hours
Insight into business policy, practice in co-
    ordinating functions, using human
    relations skills in team setting

W9. WATCH YOUR GARDEN
    GROW; 1973; UK
−; Intellect (UK) Ltd
4 to 9 years old
Board game in which players compete to
    bring their gardens into bloom first;
    gives insight into importance of seasons,
    weather and pests

W10. WATERLOO; 1962; USA
−; Avalon-Hill Inc
General
2+; 3-6 hours
Insight into Napoleon's strategy and tactics

W11. WATERLOO; USA
−; Simulations Publications Inc
General
Napoleonic warfare

W12. WEASEL WINDOWS; UK
P.E. Consulting Group; P.E. Consulting
    Group
College, management
Production control simulation on batch
    production

W13. WEATHER; USA
M. S. Gordon; Abt Associates Inc
Pre-school, elementary school
Practice in selecting suitable clothing and
    activities on the basis of visual weather
    signs

W14. WEMBLEY; UK
−; Aerial
General
3-12
Build-up to the F.A. Cup Final

W15. WFF: THE BEGINNER'S
    GAME OF MODERN
    LOGIC; 1963; USA
L. E. Allen; Wff'n Proof Inc
Primary Grades
2; 10-60 minutes
Practice recognition and construction of
    mathematical logic formulae

W16. WHAT-GOES-WITH-WHAT
    PUZZLE DOMINOES; UK
−; Abbatt Toys Ltd
5-10 years old
Practice association of ideas

W17. WHAT'S THAT ON MY
    HEAD; USA
−; Games Research Inc
General
Deductive reasoning

W18. WHAT THE LETTERS SAY;
      1955; USA
E. W. Dolch; Garrard Publishing Co
Kindergarten, grade 1
2-6; 10-15 minutes
Teaches that every letter has a name and a
   sound

W19. WHAT WOULD YOU HAVE
      DONE? 1972; UK
Schools Council; Longman
Secondary school
Designed to help adolescents get on with
   people by developing their ability to see
   other's points of view

W20. WHEELS; 1971; USA
D. Dodge, J. Powell; Macalester College
High school
Insight into budgeting for hire purchase of a
   car and one year's maintenance

W21. WHICH WAY TO CANADA?;
      1971; UK
R. Walford; British Broadcasting Corporation
10-11 years old
1+

W22. WHO GETS IT?; 1954: USA
E. W. Dolch; Garrard Publishing Co
Pre-school to grade 1
2-5; 10-15 minutes
Prepares reading readiness

W23. WHURPS; 1973; UK
C. Elgood; C. Elgood
Junior management
4-24; 2-4 hours
Teaches economics, social responsibility
   and the role of communication in
   business

W24. WIDE WORLD; USA
−; Parker Bros
General
Air travel game

W25. WILD ANIMAL PICTURE
      DOMINOES; UK
−; Waddingtons Ltd
3 years old and over
2+
Teaches number and four rules

W26. WILD LIFE; UK
−; Spears Games
10 years old to adult
3-5
Teaches world geography, animal species
   and their habitats

W27. WILD LIFE LOTTO; UK
−; Spears Games
5 years old and over
2-5

W28. WILDERNESS CAMPAIGN;
      USA
−; Simulations Publications Inc
General
American Civil War. Confederate crisis
   of 1864

W29. THE WILSON SENIOR
      HIGH PRINCIPALSHIP;
      USA
−; University Council for Educational
   Administration
Educational administrators

W30. WIN, PLACE & SHOW; USA
−; 3M Company
General
3-6
Horse racing

W31. WINE CELLAR; USA
Dynamic Design Industries
−; General
Stocking a wine-cellar

W32. WINTER WAR; USA
−; Simulations Publications Inc
−; General
Finnish campaign

W33. WOMAN AND MAN; USA
—; Psychology Today Games
Adults
Insight into sexually assigned roles in
  Western society

W34. WOMAN AND MAN; USA
—; Dynamic Design Industries
General
Psychological game on male/female roles

W35. WOMEN'S LIB; USA
—; Urban Systems Inc
Adults
4-7
Insight into the Women's Liberation
  Movement and its interaction with
  society

W36. THE WOOD BLOCKS GAME;
       1969; USA
—; Training Development Center
Management
2-16; 1-2 hours
Insight into joint goal-setting and its
  relation to the style of supervision

W37. WOODBURY POLITICAL
       SIMULATION; 1969; USA
R. C. Wood, B. Seasholes, M. H. Whithead,
  H. R. Coward; Little Brown
College
24-72; 8-9 hours
Insight into dynamics of elections

W38. WORD BINGO; 1970; UK
—; Community Service Volunteers
Immigrants
3+
Reading practice

W39. WORD COVER; USA
—; Houghton Mifflin
Pre-readers
1-10
Practice in recognition of high frequency
  words

W40. WORD DOMINOES; 1970; UK
—; Community Service Volunteers
Immigrants
Insight into the relationship of sound to
  symbol in reading

W41. WORD DRAUGHTS; 1970;
       UK
—; Community Service Volunteers
Immigrants
Practice in reading

W42. WORDFORM; UK
—; Power Play
Primary, secondary school
Vocabulary and spelling practice

W43. WORDMAKING AND WORD-
       TAKING; UK
—; Spears Games
General
Vocabulary extension and spelling
  practice

W44. WORDMASTER MAJOR; UK
M. Hardiment, J. Hicks, T. Kremer;
  MacDonald Ltd
Primary school
Reinforces and improves reading, spelling,
  word association and sentence formation

W45. WORD POWER; 1967; USA
—; Avalon-Hill Inc
8 years old and over
2-6; 1-2 hours
Practice in vocabulary building

W46. WORD ROADS; 1970; UK
—; Community Service Volunteers
Immigrants
Reading practice

W47. WORLD DIPLOMACY; 1968;
       UK
P. J. Tansey; S.A.G.S.E.T.
Secondary school, colleges of education
6
Insight into diplomatic and political
  processes

W48.   WORLD GAME; 1970; USA
R. Buckminster Fuller; Spaceship Earth
  Exploration
Junior high, high school, college
1-27; 5 day-3 months
Development of environmental planners

W49.   WORLD SERIES
       COMPUTERIZED BASEBALL;
       USA
−; E. S. Lowe
General
2
Baseball

W50.   WORLD SOCCER; UK
−; Seven Towns
General

W51.   THE WRITING GAME; 1969;
       USA
−; Training Development Center
Management
3-16; 1-2 hours
Practice in writing and reading of effective
  memos

W52.   WORLD WAR I ARMOUR;
       USA
−; Guidon Games
General
Armoured warfare

Y1.   YAHTZEE; 1972; UK
−; Parker Bros
General
Practice in number relations

Y2.   YEAR OF THE LORD; USA
−; Avalon-Hill Inc
General
Illustrates church calender

Y3.   YEAR OF THE RAT; USA
−; Simulations Publications
General
2
1972 communist spring offensive in
  Vietnam

Y4.   YES, BUT NOT HERE;
      1970; USA
J. Fischer, R. Montgomery, J. J. Zif;
  McMillan Co
Junior high, high school
20-40; 2-3 hours
Insight into formation of strategy;
  negotiation and conflict between public
  and private interests in housing problems

Y5.   THE YOUTH CULTURE
      GAME; 1971; USA
L. A. Callahan *et al.*; Urbandyne
Young people
6-12
Stimulates communication between youth
  sub-culture and mainstream US culture

Y6.   YOUTH LEADER SELECTION
      SIMULATION; 1969; UK
−; Methuen Ltd
College, management
Insight into problems of interviewing and
  assessing

Z1.   ZERO; USA
−; Simulations Publications Inc
General
Insight into tactical air warfare in 1941-45

Z2.   ZIGGURAT; UK
−; Mosesson Games Ltd
General
2-6
Gives practice in use of spatial concepts
  and arithmetic

Z3.   ZOUPA; UK
P.E. Consulting Group; P.E. Consulting
  Group
College, management
2+; 12 hours
Practice in marketing management for
  industrial and consumer products

5.2 Subject index

Advertising
A9, A11

Aid and disaster relief
A23, A24, C86, D30, D31, D32, D45, E81, H18

Colour recognition and mixing
A46, B45, C76, C77, C103, C142, D1, D51, H32, M11, M69, M107, M110, P96, P123, P132, R9, S58, S59, S110

Communication
General
B5, B50, C70, C73, C83, D79, E62, E67, G5, G6, H18, I20, J8, L21, L39, R5, W19, W23, W51, Y5
Grammar and syntax
L9, L10, L11, L13, L14, N37, S39, S40, S71, S184, S197
Phonics and word formation
A48, B32, B40, B61, B86, C17, C116, C117, C153, C154, D40, D41, D59, F4, F7, F11, F66, G39, G62, H9, M62, M121, M145, M146, M158, O9, P13, P39, P40, P46, S17, S46, S60, S121, S135, S136, S152, S154, S171, S219, S220, T15, T74, W18, W40
Reading
B38, C5, C36, D40, D41, D59, D64, D65, D66, D67, F3, F4, F11, G31, G62, G63, H8, I85, J14, K2, K4, L27, L37, L38, L49, R12, R13, R14, R24, R25, S126, T105, V14, V15, W22, W38, W39, W40, W41, W44, W46, W51
Spelling
A10, A79, B86, C42, C154, L28, L37, M61, M121, O44, P65, R54, R55, S22, S24, S46, S145, S146, S147, S148, S149, S151, S176, T20, T110, W42, W43, W44
Vocabulary
A10, A79, B30, B86, C5, C42, C159, F40, H8, H22, L12, L28, M12, M60, M111, M157, P3, P24, P65, P100, R55, S22, S23, S40, S63, S145, S149, S151, T110, V5, W42, W43, W45

Computing
B26, B65, C15, C26, C34, C96, C97, C98, F10, F21, F22, F23, F24, I66, M152, P114, S11, S84

Counselling and careers
A76, C21

Decision making
A23, A24, A37, A62, A68, A77, B3, B73, B74, B77, B79, C7, C15, C58, C148, D2, D4, D5, D9, D10, D11, D12, D13, D14, D15, D16, D17, D18, D27, D44, D79, D80, E42, E55, E58, E68, F26, F51, F53, G65, H7, H18, I3, I8, I59, I64, I78, J6, J16, L7, L17, L18, L19, L21, M22, M23, M24, M29, M31, M32, M39, M42, M55, M56, M116, M138, N41, O8, O50, P1, P12, P48, P50, P73, P86, Q2, R4, R34, S74, S80, S86, S87, S108, S133, S161, S164, S203, S205, S208, T72, T73, T77, T80, T82, T103, U4

Drugs
C165, D73, D74, D75, E29, M44

Economics
C52, C53, C54, C60, C61, C70, C119, C120, C133, C136, C150, D16, D17, E8, E9, E10, E11, E12, E17, E75, E81, E87, F8, F28, F34, F35, F62, F64, G68, I71, K15, M45, M46, M47, M48, M50, M142, N5, N6, N16, N42, O55, P7, P50, P76, P129, P130, S43, S167, S168, S169, T89, W23

Education and training
A1, A69, A70, C16, C73, C74, C85, C100, C127, D13, E14, E15, E16, E18, E19, E27, E28, G2, G9, H15, H16, H26, H37, I7, I23, I24, I25, I26, I27, I28, I29, I30, I31, I32, I33, I34, I35, I36, I37, I38, I39, I40, I41, I42, I43, I44, I45, I46, I47, I48, I49, I50, I51, I52, I53, I54, I55, I56, I62, J1, J2, L2, L18, L23, L32, L33, L46, L53, M4, M5, M6, M7, M8, M125, M143, M149, O59, P70, P112, R29, R38, R50, S26, S30, S31, S32, S35, S52, S77, S79, S109, S223, T21, T22, T23, T24, T25, T26, T27, T28, T29, T30, T31, T32, T33, T34, T35, T36, T37, T38, T39, T40, T41, T42, T43, T44, T45, T46, T47, T48, T49, T50, T51, T52, T53, T54, T81, T93, T94, U8, W29

Finance
Accounting
A3, A7, P128
Currency handling
C38, D7, D8, D69, F5, M131, N18, P16, S64, S162, S199
Investment and insurance
A5, I3, S85, S142, S178, S179, S180, S181, S182, S183, T90, T95, T96, U9

Geography

Africa
A23, A24, C148, D63, G38, K5, M50, P76, S69, S94, S102

Asia
A21, A26, D2, D63, E30, G64, K19, K20, K21, K22, M130, N35, P15, P20, S91, S92, S100, S101, V8, V9, V10, V11, Y3

Australasia
A26, F25, G66, G67, R2, T87

Europe
B14, B16, B51, B53, B54, B59, B68, C3, C10, C145, D6, D19, D77, D78, E22, E52, F46, F59, F61, F63, G29, G37, G49, G51, I68, I86, L25, M127, N1, N2, N3, N31, N32, N33, N34, N40, O51, O57, P11, R23, R30, S4, S88, S95, S97, S103, S140, S141, T18, T107, U4, U5, V7, W32

General
A2, A39, C1, C34, E52, E84, E85, E86, F63, G52, K14, L42, L43, L45, M114, N42, S69, W21

Middle East
C66, C102, M29, S89, S98

North American and the Arctic
A34, A39, B15, B46, C13, C24, C25, C33, C39, C43, C75, C112, C113, C129, C136, D27, D50, E32, E37, E85, F64, G34, G35, H28, I11, I68, I78, K14, L32, M133, N9, N10, N12, N26, N39, P7, P48, P74, P84, P85, P93, P97, Q4, R49, S48, S49, S51, S82, S130, T76, W21, W28

Russia
B6, B17, B18, C3, D28, E1, K24, M147, N9, O60, P8, S164, T18, W3

South America
A43, C23, C34, C132, D27, D33, E86, I68, L6, P5, P6, P72, S90, S93, S99

Government, politics and diplomacy

Diplomacy
A2, A72, A80, B52, B57, C3, C66, C67, C68, C69, C102, C108, C147, C148, C149, C162, D2, D30, D35, D36, D42, D43, F47, F51, G8, G11, G12, G29, G51, H40, I11, I71, I72, I73, I75, I76, K17, K23, L24, M1, M35, M122, M159, M160, N9, O57, P9, P69, P73, P97, S20, S75, S81, S87, S88, S89, S90, S91, S92, S93, S94, S95, S97, S98, S99, S100, S101, S102, S103, S104, T59, T60, T67, T112, W6, W47

Local government
A29, A39, B66, C9, C51, D23, D24, D25, D57, E32, I1, I4, I14, I21, I60, L41, L44, M116, M117, M118, M129, M151, P101, S33, S111, S209

National government
A34, A35, C13, C39, C112, C113, G7, G18, G68, I67, U4, U5

Politico-military
A21, I71, M130, M137

Politics
A34, A63, B37, B52, C2, C3, C13, C39, C52, C53, C54, C70, C102, C108, C109, C124, C136, C147, C148, C149, C166, D2, D20, D22, D24, D27, D54, D63, D71, E24, E25, E26, E30, E31, E81, F64, G14, G27, G30, G68, I59, I71, I72, I73, I75, I78, L29, L30, L31, M1, M35, M102, M132, N4, N26, N39, P7, P11, P62, P71, P72, P74, P75, P76, P88, P91, P93, P94, P95, P122, R26, R36, S49, S82, S83, S172, S200, S218, V13, W35, W37, W47

Pressure groups
A28, C51, P118, P124, S111

Health, welfare and medical
A17, B36, B37, D56, H16, H27, H30, H34, H35, I18, M104, M108, S72, S73

History

Air Force
F36, L51, Z1

General
A14, A30, A34, A35, A39, A72, B56, C75, D34, D50, D71, F59, F64, G29, H10, H14, H28, I78, I85, J10, K6, L3, L30, M34, M133, N10, P117, R4, R40, R49, S1, S2, S20, S48, S69, S113, S174, S196, T91, T102, V7

Military
A6, A19, A20, A30, A31, A38, A54, A55, A64, A65, B6, B11, B12, B13, B14, B15, B16, B17, B18, B42, B46, B51, B53, B54, B68, C10, C31, C33, C88, C145, D3, D6, D19, D28, D77, D78, E1, E22, F15, F46, F60, G34, G35, G49, G55, G64, G66, G67, I82, I83, K19, K20, K21, K22, K24, L25, M57, M126, M147, M156, N1, N2, N3, N31, N32, N33, N34, N35, N40, O45, O51, O60, P2, P8, P15, P20, P37, Q4, R30, R33, S4, S47, S68, S131, S140, S141, S164, S186, S187, S192, T18, T71, W5, W7, W10, W11, W28, W32, W52, Y3, Z1

Naval
B35, F18, F33, J16, S61, T92, T107, U14

Housing and urban development
B37, H31, I22, I60, L33, M15, M115, N12,
N23, O54, P84, S51, S77, T58, T88, U13, Y4

Human behavioural studies
Anthropology
B72, C23, C24, C25, C132, C164, D39, D72,
F25, H38, I11, I81, P85, P87, S130, T76,
T87
Group behaviour
A37, B41, B70, C11, C71, C122, C165, D5,
D14, D75, E65, E71, E74, E76, E77, F50,
G4, G6, G17, H36, H37, I57, I61, I70, J6,
J9, P82, S108, T64
Psychology
B69, C110, F6, I61, J17, P118, R44, W33,
W34
Sociology
A14, A42, G48

Industry and commerce
Agriculture, horticulture and forestry
C34, F8, F9, F53, F55, G13, G19, H19,
M135, N41, P86, P126, P127, P128, P129,
S74, S203, W9
Aircraft and aero-space
A16, A25, A26, A27, A28, B60, I69, M3,
M141, S21, S38, T68, W24
Automobile industry
A68, B23, C141, D5, F52, L26, L43
Banking
B3, B4, B5, F14, P103, S165
Chemical industry
C27
Construction industry
E75, E89, E90, O55
Engineering
B75, D58, G54, M29, M31, M39
Entertainment industry
C35, C50, M150, Q7
Fire service
F53
Hotel and catering industry
C133, C134, P47, W31
Iron and steel industry
I14, I84, T82
Journalism
A80, D57, F65
Mining industry
B74, C147, H10, R2, R49, S174

Oil industry
A68, E45, E46, E47, E48, E49, N42, N43,
O6, O8
Policework and criminology
C56, C59, D29, G45, I19, J17, L15, R6, S41,
S76, S77, S114
Trading
B22, B71, E31, I6, I77, M140, P87, S65, T89,
T91, T102, W20
Transport industry
B71, B73, B74, C1, C14, C53, C54, C136,
D52, G41, G52, I65, L7, L45, P1, P117, R7,
R8, S61, S77, S78, S61, S77, S78, T68, T73,
T97, T98, T99, T101, W24

Languages
A45, C37, C65, L8, S122

Law
A8, D23, G48, G53, H14, L16, S86, V6

Logic
A22, A56, C47, C130, C138, D29, E21, E36,
F61, K3, K8, K9, K10, K11, K12, K13, K18,
N30, N36, P17, P90, Q5, R3, S53, W15,
W17

Logistics
A19, A20, C7, K23, N15, N33, S163, T62

Management
Administration
A1, A12, C71
Advertising
A9, A11, B82, M42, M105, P61
Financial
A66, B2, B3, B4, B5, B63, B64, B73, B75,
B77, B82, B84, C15, C26, C30, C82, C84,
C144, C160, D46, E42, E49, E69, E82, F16,
F19, F26, G43, I3, I8, I15, M63, M90, M109,
P49, P52, P53, P54, P55, P56, P57, P58, P59,
P98, P125, R32, S28, S214

General
A5, A7, A16, A36, A44, A61, B23, B27, B49,
B58, B76, B78, B79, B80, B81, C27, C89,
C131, D9, D11, D21, E46, E47, E48, E49,
E51, E54, E55, E56, E60, E70, E72, E73,
E78, F55, G24, G25, G28, G47, G54, H11,
H12, H15, H25, H30, I2, I3, I8, I10, I13,
I15, I64, I68, M19, M20, M21, M22, M23,
M24, M25, M26, M30, M31, M32, M49, N8,
N14, O1, O4, O7, P35, P38, P99, P103,
P104, P126, P127, P129, P130, Q2, Q3, S26,
S74, S78, S116, S117, S118, S158, S161,
S175, T75, T77, T78, T80, U2, U10, U11,
U12, V3, V4, V12, W23, W36

Maintenance
C12, C22, P51

Marketing and sales
B73, B75, C98, C101, C160, D16, D18, E57,
E69, E80, E82, E87, F2, F26, F67, G22,
I17, L48, M17, M18, M27, M51, M52, M53,
M54, M55, M56, M59, M105, M136, N17,
P47, P79, R39, R41, S3, S5, S6, S7, S8, S9,
S10, S37, S85, S205, S208, S215, U6, Z3

Negotiations
C126, C128, E50, E71, G56, G57, H2, H3,
H4, N9, N10, N11, P70, P78, P111, S45,
S172, Y4

Personnel
A52, A60, A61, A67, C28, C29, C71, C126,
C128, D11, D14, D17, D47, D48, D49, E32,
E50, E61, E66, E67, E89, F1, G20, G22, G23,
G56, G57, H2, H3, H4, H10, H14, H36, H37,
I9, I12, I16, I79, J7, J8, L3, M25, M28, M33,
M37, M148, N11, P27, P28, P29, P30, P31,
P32, P33, P34, P92, P111, S36, S45, S167,
S177, S189, S194, S195, S206, S211, S212,
S213, T79, U8, W8, Y6

Policy
A16, B80, E59, G19, I8, I67

Production
B73, B75, C6, D15, D17, D46, D47, D58,
E57, E69, E82, E91, F67, J6, M39, M40,
M41, M42, M124, N1, O56, P60, P79, P105,
P107, P108, P109, P110, P119, P120, S125,
T55, T72, T103, U7, W12

Resources
B84, C4, C7, C124, F67, M27, M64, M65,
M66, N15, O50, P106, R2, R37, S16, S166,
T88

Strategy
See page 189

Mathematics
Addition
A15, B83, C57, C139, D69, E41, F30, M9,
M13, M67, M86, N49, O46, P4, S12, S123,
T19, T84, W1, W25

Division
A15, C57, D69, E41, F23, M67, N49, S13,
S123, W1, W25

General
A15, A61, A77, B10, B31, C143, C155,
C156, C157, D55, D56, E38, E39, E40, E41,
E63, E64, E65, F5, F22, F24, F31, F48,
F58, F68, F69, G32, G33, G46, M13, M68,
M69, M70, M71, M72, M73, M74, M75, M76,
M77, M78, M79, M80, M81, M82, M83, M84,
M85, M86, M87, M88, M89, M134, M153, N50
O10, O11, O12, O13, O14, O15, O16, O17,
O18, O19, O20, O21, O22, O23, O24, O25,
O26, O27, O28, O29, O30, O31, O32, O33,
O34, O35, O36, O37, O38, O39, O40, O41,
O42, O43, R15, R16, R17, R18, R19, R20,
R21, S155, S201, S202, T63, T100, T109,
T116, Z2

Multiplication
A15, C57, D69, E41, F21, M67, M68, M154,
N49, P14, S14, S123, S124, S207, W1, W25

Numbers
A15, B9, C57, C78, C79, C143, D8, D62,
D69, E41, F29, J5, J13, L4, L5, L49, M67,
M74, M77, M78, M79, M80, M82, M83, M84,
M85, M112, N46, N47, N48, N49, O47, O48,
O49, P64, P96, R9, R51, S156, T59, T84,
W1, Y1

Spatial relationships
A33, A40, A41, A47, B43, B45, B85, C40,
C104, C105, C106, C107, C114, C142, C161,
D51, H20, H21, J3, L1, M11, M43, M91,
M92, M93, M94, M95, M96, M97, M98, M99,
M100, M101, M107, M110, M155, O2, P22,
P23, P25, P26, P123, P132, R45, R46, S44,
S56, S57, S59, S110, S210, S221, T61, T70,
T113, Z2

Subtraction
A15, B83, C57, C139, D69, E41, M67, M87,
N49, O46, S15, S123, T19, W1, W25

Topology
B24, K16, Q6

Miscellaneous
A75, B39, B62, C8, C19, C43, C44, C46, C49, C62, C63, C99, C118, C121, C125, C137, C151, C152, D37, D38, D58, D68, D70, D81, E83, F3, F20, F32, F41, F42, F70, F71, G3, G10, G16, G36, G40, G51, M13, M23, M24, I62, J4, J11, K1, L29, L32, L40, L52, M2, M10, M58, M103, M106, M128, M144, N44, O61, P10, P21, P36, P41, P42, P43, P44, P45, P63, P80, P81, P83, P121, P131, P133, Q1, Q8, R1, R22, R28, R31, R42, R53, R56, S18, S27, S29, S34, S50, S54, S55, S62, S66, S70, S105, S106, S107, S120, S132, S137, S139, S144, S173, S185, S190, S191, S222, T56, T85, T104, T108, T114, V1, W2, W4, W13, W16, Y2

Motoring and road safety
R43, R47, R48, S143, T111

Music
D60, D61

Natural history
Animal recognition
A45, A46, A47, A48, A49, A50, A51, J12, W25, W26, W27
Ecology
B1, E2, E3, E4, E5, E6, E7, E13, E88, N36, P89, R26, S217
Environment
E34, E35, I81, M16, P50, W48
General
C115, E3, E4, E5, E6, E7, E20, F12, G1, G15, H34, H35, M14, P129, W26, W27
Pollution
A57, B48, C81, C115, D44, P76, P77, P78, S77, S115, S119, S170, S216

Planning
A28, A29, B3, B36, B58, B67, C51, C100, D11, D16, D46, D48, E16, E19, E26, E45, E58, E74, F1, F2, G22, G44, G65, H16, H27, I1, I14, I21, I22, I80, L15, L51, M15, M38, M115, N14, N19, N20, N21, N22, O54, P48, P61, P68, P94, P95, P103, P104, P115, P126, P128, Q3, R37, S26, S127, S159, S160, W48

Problem-solving
A69, A70, A71, B36, C55, C97, C150, G4, G6, G58, G60, G61, H26, I70, L21, P51, P102, P107, Q3, R10, R11, T21

Race relations
B37, E37, H5, S204

Religious and ethical topics
A64, A74, I5, J10, L34, L35, L36

Science
Chemistry
C41, F54, M139
General
D4, H34, H35, I58, Q5, S19, S138
Physics
C48, E23, N25, S42, S138

Simulation
M22, U7

Sport
American football
C72, C92, F45, P113, R52, S25, T65
Association football
B28, B44, C123, F43, F44, F57, I74, L22, O3, P18, P19, S128, S129, T5, T6, T7, T13, W14, W50
Baseball
B7, B8, B24, C90, C91, S112, W49
Basketball
P66,
Cricket
C20, C146, T1, T2, T3
Golf
C32, C93, H1, T66
Hockey
C94
Horse racing
E43, T83, W30
Ice-hockey
B47
Rugby football
T4
Sailing
C95, H17, R27

Strategy
Air Force
L51
General
A58, A59, A63, C24, C25, C30, C45, C69, C80, C87, C111, C148, C163, E9, F13, F45, F56, G14, G27, G42, H39, I1, I22, K7, L26, L48, M123, N29, O58, P65. P123, S34, S67, S134, S200, T12, T14, T16, T86, T88, T106, Y4
Management
B80, C15, C134, E71, E79, G20, I8, I80, M26, M53, M55, P49, P68, R34

Military
B11, B18, B19, B42, B59, C10, C64, C135,
C140, D6, F15, F60, G34, G67, H6, K23,
M126, N13, N45, S163, T9, T10, W5, W10,
W11
Naval
B33, B34, J16, M36, N24, T92

Systems
N27

Tactics
Air Force
A18, A78, B12, B13, F17, F36, F37, F38,
S157, S198, Z1
General
A32, A68, B21, C18, C42, C80, E53, G21,
J15, K7, M55, M123, N28, O5, P49, P67,
T16, T70, T115, V2

Military
A6, A19, A73, B11, B18, B20, B42, C135,
C140, D3, F27, F60, G64, H6, H33, K23, L6,
L47, M147, N15, N33, N45, O45, P8, P15,
P20, P37, P116, R23, S47, S163, T8, T9,
T10, T11, T17, W10
Naval
A13, B33, B55, D76, F18, F33, J16, M36,
N24, T92, T107, U1

Youth service
A4, A53, A67, C58, D10, D26, E44, F3, F4,
I5, J9, L17, L19, L20, M30, M113, M138,
O52, O53, Y5, Y6

5.3 Author index

Abelson, R. P., C13
Abrimes, S., S189
Abt, C. C., B5, E16, E31, G50, T57, R6, S77
Adair, C. H., E83
Adams, C., N25
Addis, M., S203
Adolph, P., C123, I74, T5, T6
Ahern, J. F., R38
Aitchison, J., D56
Aldrich, A., L20, M30, O53
Allen, L. E., A15, A22, A56, C47, C130,
    C138, E21, E36, E41, F61, K3, K8, K9,
    K10, K11, K12, K13, K18, N30, N36, O10,
    O11, O12, O13, O14, O15, O16, O17,
    O18, O19, O20, O21, O22, O23, O24, O25,
    O26, O27, O28, O29, O30, O31, O32, O33,
    O34, O35, O36, O37, O38, O39, O40,
    O41, O42, P17, Q5, R3, R15, R16, R17,
    R18, R19, S53, W15
Allen, R., A39, B9, B10, C143, E52, E83, I58,
    P118, R24, R25, R43,
Alpert, S., G48
Ameen, D. A., C4
Anderson, D., A69, M8
Anderson, J., D44, S119
Andlinger, G. R., B78
Archey, W., M33, S10
Armillas, I., G18

Armstrong, R. H. R., A29, C124, D31, E33,
    E51, E81, G44, I21, L41, M21, N27, P115,
    S107
Arnett, J., C6
Atherton, R., S11
Atkins, T., E15
Atthill, C. R. A., D16, D17, D18
Attworth, E. R., T75
Audlinger, G. R., H11
Austwick, G., H21
Ayal, I., M55, S208
Aydon, C. N., F26
Ayers, M., S59

Babb, E. M., P126, P127, P128, P130
Babbidge, H., C129
Banks, M. H., C102
Barber, D., A67
Bare, B. B., P129
Barnett, H. L., M129
Barringer, R., S51
Barron, T. W., S96, S104
Barry, W. S., B60
Barton, R. F., I2
Bass, B. M., E61, E62, E63, E64, E65, E66,
    E68, E70, E71, E72, E74, E76, E77,
    E78, U12
Bauer, C. S., F27

Bayley, J., S108
Bazelon, J., C155, C156, C157, C158
Beckett, B., I16, P20
Belkin, B., F12
Benor, D., S26
Benson, D., S75
Benson, D. K., I71
Benyenuto, A., M7
Berne, E., G17
Betts, P. W., R11
Blackshaw, R., C67
Blake, A. F., S145
Blaxall, J., M34
Bloomer, J., C48
Bloomfield, L. P., G128, M137
Boardman, R., M35
Boguslaw, R., P62, S175
Bolton, D. R., M68, S35, T22
Boocock, S. S., G27, L32, P10
Bonacaich, E., I67, S26
Bornstein, S., T99
Bracken, I., N6
Bradford, G., H33
Bradley, M., S200
Bradley, R., A30
Branfon, R., E31
Brent, D., S109
Brett, J., T110
Brett, P., T110
Broadbent, F. W., L23
Brown, L., I70
Brown, R., V5
Brown, W. A., F27
Bruce, D., S195
Bruna, D., C49, D38
Buck, L., O1
Buckminster Fuller, R., W48
Buffie, E. G., H37
Bunker, R. M., S92, S101
Burgess, P. M., I59, I71, S83
Burnett, J. D., F55
Butchers, R. J., C56

Calder, N., E34
Calhammer, A., D42, H39
Callaghan, D., J17
Callahan, L. A., E14, U13, Y5
Calland, A. M15
Campbell. F. M., T75
Campbell, M. E., S94, S102
Carsello, C. J., B32, D40, D41, F66, O9, S152, S154, S219, V14

Chapman, K., M46
Chase, R. M., C133, C134
Cherryholmes, C. H., I72, I73, R35
Churchill, G., J6
Cinquegrama, A. R., S95, S103
Clark, C., S116
Clark, J. M., E52
Clarke, M. F. C. A23, A24, D33
Clarke, R., C148
Clary, B., S26
Clayton, M., G7
Cocanower, N., K2
Cohen, K. C., C27, D75
Cole, J. P., C34, E25, E26, G30, L42, L43, L44, M1, M3, O6, P47, P94, P95, R28, S160, T82
Coleman, J. S., D24, D25, E10, T14
Coller, H. R., C76
Colomb, P., D76
Conway. J., G16
Cooke, J. E. E89, P111
Cooper, G., T58
Cooper, J. R., D46, D47
Copeland, M., G14
Coplin, W. D., A34, A35, B63, B64, C112, C113, D12, F51, G48, P93, P97, S173
Cormier, C. R. S91, S100
Coward, H. R., W37
Cranmer, H. J., B4, C84, F28, M45, N5
Craven, J. A., C89
Crawford, J., T102
Crawford, R. J., B60
Crean, P., H31
Crick, E. B., P62
Cross, F. L., S80
Cruickshank, D. R. I23, I24, I25, I26, I27, I28, I29, I30, I31, I32, I33, I34, I35, I36, I37, I38, I39, I40, I41, I42, I43, I44, I45, I46, I47, I48, I49, I50, I51, I52, I53, I54, I55, I56, T24, T25, T26, T27, T28, T29, T30, T31, T32, T33, T34, T35, T36, T37, T38, T39, T40, T41, T42, T43, T44, T45, T46, T47, T48, T49, T50, T51, T52, T53, T54
Curtis, S., R56
Cuthbert, N., D14
Cuthbertson, K., P20
Cyros, K. L., I62

Dale, A., S116
Dal Porto, D., B57, C39, H14, R4, S140
Danten, C. A., W8

Darden, B. R., O56
Darrow, C. W., M142
Daubney, A. C., C58
Davies, R. H., B81, P62
Davis, F. C., A2, A72, G29, S139
Davis, J., O1
Davison, W. P., P124
De Bono, E., L1
Deeley, T., P49
Deep, S. D., E77
De Koch, P., D27, D54, M130, P7, S204
Del Solar, D., P72
De Ridderstad, T17
De Sola, C13
Dickson, D. L., G67
Dill, W., C27
Dodge, D., M151, W20
Dolch, E. W., C117, F29, G62, G63, H8, K14,
    M13, M60, P16, P45, R12, S12, S13, S14,
    S15, S220, T15, T59, V15, W18, W22
Dorwart, H. L., C104, C105, C106, C107
Downes, L. W., F30, P14, S123, S124, S156
Duke, R. D., A57, C81, M115, M116, M117
Dunnigan, J., O61, U14
Durham, R., H10, T102
Durham, V., H10

Eglinton, G., C41
Eisgruber, L. M., P126, P127, P128, P130
Eldon, B., E59
Elgood, C., F67, W23
Elliot, J., C151
Elliot, R. W., E67
Ernst, M., S51
Erskine, R., A70
Estey, E. E., I64
Etzioni, A., R37

Fain, J., N35
Fain, W., N35
Fairhead, J. N., E59, M54, P34, S212, T78
Farrar, G., B36
Faubert, R., A30
Feig, W., S49
Feldt, A. G., C60
Ferguson, S., T81
Fertig, P., B77
Field, C., G7
Finlay, J., I1
Finseth, K., P48
Fischer, J., Y4
Fischer, S. J., I22

Fischer, S. S., S172, S194
Fitzgerald, B. P., I84
Fogler, H. R., P49
Folks, I. T., S215
Folks, W. R., S205, S215
Forbes, J., C74
Ford, M., G3, J4, P80
Frey, W., F19
Fuller, E. I., N41, P86

Gamson, W. H., S79
Garland, K., C114, F32
Garvey, G. M., S82
Gearon, J. D., L3, M102, W6
Gibbs, G. I., S19
Gideon, V. C., P52, P53, P54, P55, P56
    P57, P58, P59
Giffin, S. F., C149
Gill, J., B25
Gill, S., F21, F22, F23, F24
Gillespie, P. S., R50
Gilner, E., P132
Glazier, R. E., B72, C164, E16, G50, H18,
    R6, S45, S77
Glover, C. M., F58, G32, G33
Glynn, D., D64, D65, D66, D67
Goetze, R., G7
Goldhamer, H., C69
Gollay, E., M2
Goodman, F., T64
Goodman, R. F., I67, P70, S26
Goodman, W., S69
Gordon, A. K., C75
Gordon, M. S., C77, C136, P72, S56, W13
Gotkin, L. G., C76, L9, L10, L11, L12, L13,
    L14, L38, M69, M70, M71, M72, M73,
    M74, M75, M76, M77, M78, M79, M80,
    M81, M82, M83, M84, M85, M86, M87,
    M88, M91, M92, M93, M94, M95, M96,
    M97, M98, M99, M100, M101, P22, P23,
    P24, P25, P26, T105
Gottheil, E., E16
Grant, C., A6, B15, B16
Graves, R. L., I80
Gray, C. F., M24
Greene, L., P118
Greene, J. R., H11, I17, M49, M64, R39,
    T80
Greene, R. M., S150
Greenlaw, P. S., D5, F19, M56, P119
Grierson. J. M., B59
Groff, W. H., P114

Groom, A. J. R., C102
Gruen, C., S51
Guetzkow, H., I72, I73
Gunnell, J. B., U8
Gygax, G., D78

Haas, E66
Hack, W., A69
Hagstrom, P., O1
Halmstad, D. G., S85
Hanan, M., A9, A11, M17, M51, M53, P105, S5, S6
Hanneman, G. J., S106
Harader, W., D20
Hardy, J., P101
Hargreaves, S., E45
Harlan, J. A., A62
Harris, B., A70
Harris, J. R., E10
Harrison, M. L., A48, B61, D59, P39, P46, S40, S71, S184
Hartley, W. C. F., I8
Hayes, D., E69
Hayes, J., S144
Hayes, M., S127, W5
Heap, J. L., S217
Heier, W. D., E60
Heilman, A. W., B32, B40, D40, D41, F66, O9, S152, S154, S219, V14
Hein, P., A41, B45, C161, H20, N29, T113
Helmkamp. R., B32, B40, D40, D41, F66, O9, S152, S154, S219, V14
Hemming, C., D43, K17
Hendricks, F., G65, P68, S51
Henshaw, R. C., E58
Henston, A., M141
Herrick, C. S., S86
Herron, L. W., E55
Hester, J., G7
Hicks, B., T104
Hicks, H., T104
Hicks, J., C139, H9, S57, W44
Hiersteiner, R., S172
Higgins, H. E., E44
Hills, R., H25
Hindmarch, N., M158
Hinings, P. H., M105
Hinman, H., G19
Hobson, B., S144
Hobson, C., S144
Hobson, M., A29, C124, D31, E33, E51, E81, G44, M21, N27, P115, S107

Hodder, J. C., C136, S205
Hodgetts, R. M., T79
Hoheisel, W., G8
Hoinville, G., E35
Holland, L. V., S39
Hollister, R., G7
Holtenstein, M. P., P119
Hoolim, H., P21
Hopkins, T. L., S89, S98
Horn, R. E., E32, F13, P12, P60, S214
Horrocks, B., C80
Horton, I., S158
Horvat, J., P112
Howells, L. T., I80, P34
Huang, P. C., D81
Hudgell, B. A., S43
Hull., A., S43
Hunter, G. T., M22
Hunter, J. R., E51
Hutton, R., G19

Ighodaro, A., G3, J4, P80
Immegart, G. L., M7, S109
Inbar, M., C86
Irving, W. S., V1
Isber, C., C164
Istvan, D. F., B77

Jackson, J. R., E58, S114
Jackson, K., T93
Jacobs, L., T73
Jaffee, C. L., P104
Jeffrey, G., D35, D36
Jenkins, J. R. G., P61
Johnson, C., D78
Johnson, H. E., M124, U7
Johnson, L., C85
Johnson, R. R., D4
Jones, J. K., A8, A28, A60, A80, D57, F65, R5, R22, S62

Kaplan, A., F64, P117, R20
Kaplan, H., D30
Kaunfer, M., S194
Kell, W., A7
Ketner, W. D., C6, C26
Kibler, F. N., S90, S99
Kinley, H., C24, C25, S27
Kirlin, J. F., I67
Klapfer, L. E., I58
Klasson, C., S116
Klietsch, R. C., I4

Kniffin, F. W., M56
Krawitz, M. J., O4
Kremer, T., C139, H9, S57, W44
Kriebal, C. H., F2
Kuch, T., M160
Kuehn, A., C27
Kugel, I. P., O10, O11, O12, O13, O14, O15, O16, O17, O18, O19, O20, O21, O22, O23, O24, O25, O26, O27, O28, O29, O30, O31, O32, O33, O34, O35, O36, O37, O38, O39, O40, O41, O42, Q5

La Brie, H., I70
Lackman, A. A., A25
Lacome, M., R47
Lamb, P., B86, C5, C17, C153, C154, C159, F11, G39, L37, L49, M61, M121, N37, P13, S17, S46, S60, S121, S136, S146, S149, S153, S171, S176, T74
Lane, C., A20
Langell, J. A., I62
Larson, D. L., S87
Lassiere, A., E35
Laughlin, H., M8, M125
Laulicht, J., V11
Lawson, B. R., N19, N20, N21, N22
Lebling, D., L24
Leiberman, H. R., Q3
Leonard, J. M., F62
Le Vaux, J., E32
Levin, G., C20
Levy, R., G58, G59, G60, G61
Lewis, F., P43
Life, A., E59
Lineham, T. E., D71, E20, R44, V1
Link, E79
Lipetsky, J., D39
Liss, R. L., I58
Little, A. D., S51
Livingston, S., T89
Loghan, J. P., A62
Long, B. E., R44
Longley, C., T94
Longmire, R. M., C56, L15
Loomis, R., B19, N44
Lorenzana, M. B., M9
Loughanz, J. W., A76
Loveluck, C., E82, S117, S125
Lowry, D., D78
Lucas, W. H., O56
Lunstedt, R., C39, S140
Lymberopoulos, P. J., S116

Mackie, A. D., F58, G32, G33
MacLean, C., R47
Maddison, R. N., L52
Malcolm, R. E., A3
Mallen, G., S76
Mardney, P. A., Q7
Martin, J., V11
Mason, S., L38
Mastrude, R. G., D2
Mather, P. M., C34, O6, P47
Margerison, C., F50
Maxwell, J. R., C41
May, F., S116
Mazer, E. E., M68
McAuliffe, A. C., B14
McCallum, J. C., C162
McClusky, C. W., M126
McCormick, J., U4, U5
McCosh, A., A7
McDonough, A., S118
McElhave, R. W., J11
McFarlan, F. W., M26
McGrath, M. J., C6, C12
McGregor, J., T78
McIntyre, K. F., L2, S52
McKee, P., A48, B61, D59, G31, P39, P46, S40, S71, S184
McKenney, J. L., M26
McKinnell, H. A., S166
McMillan, P61
McNair, D. D., B2, C82, M42, S47
McQueeney, W., C148
Meehan, R., L23
Meeker, R. J., S26, T60
Mette, H. H., T23
Mill, S., P97
Miller, D., M122, M144, P9, S20, T112
Miller, P. S., E16, E66
Miller, S. J., R54
Million, E. Z., O8
Mills, V. J., B35
Mitchell, C. R., S108
Moheisel, W., I11
Molan, R. E., S93
Moncrieff, B., S203
Montgomery, R. A., I22, S172, S194, Y4
Moody, R., D44, S119
Moore, M. K., H18, R6
Moore, S., A42
Mottice, H. J., B77
Murfin, M., C155, C156, C157, C158

Navin, T. R., A62
Nesbit, W. A., G68
Newman, F. M., R7
Newman, R. W., G20, G21, G22, G23
Newmark, P., S216
Nicholson, M., C66
Niland. P., W8
Noël, R. C., P71
Noon, H. R., S28

Ohm, R., E27, E28, L18, S30, S31
Olafson, H. C., J17,
O'Leary, M. K., P97
Oliver, D. W., R7
Olson, D. R., M43
O'Neill, J., Q6
Oppenheim, A. N., C102
Orbach, E., M55, P32, R34, S10, S208
Orcar, L., M46
Osgood, C. E., S34
Oswald, J. M., M159
Otlewski, R. E., C126
Owen, M., O10, O11, O12, O13, O14, O15,
    O16, O17, O18, O19, O20, O21, O22,
    O23, O24, O25, O26, O27, O28, O29,
    O30, O31, O32, O33, O34, O35, O36,
    O37, O38, O39, O40, O41, O42

Packer, A., E9
Paling, D., F30, P14, S123, S124, S156
Park, W. R., L48
Parker, D. H., R14
Parker, J., P122
Parton, V. R., B43
Paterson, J., E50
Patterson, H., M123
Patterson, T. T., C28, C29
Pazer, H. L., P35
Pelton, W., S175
Pennington, A. J., B66
Pepper, R., M15
Peterson, L. E., I71
Pfeffer, J., P49
Pisano, J., M159
Pollard, H. R., H30
Pool, I., C13
Poplin, S., C13
Powell, J., W20
Prendergast, M. L., E20
Pugh, D. S., E59, M54, P34, S212, T78
Pyke, J. L., S88, S97

Rackham, N., N11
Rader, W., M46
Radice, H., F46
Rae, J., G4, G5, G6
Rae, R. W., L15
Ramsing, K. D., I13
Rasmussen, F. A., P78
Rath, J. J., I15
Rausch, E., B4, C70, C71, C84, D11, E42,
    F28, H2, H3, H4, I77, I79, L21, M45,
    N5, P30, P33, P106, P107, P125, S7, S9,
    S16, S213
Reese, J., A43, P5
Relf, W., A74
Renshaw, J. R., M141
Renwick, M., T91
Reske, J. D., C81
Richardson, E., L38, T105
Rints, N., E69
Roach, D. G., D71
Roberts, J., S52
Robinson, I., S51
Root, E., F7, I85, M12, M62, M145, P3
Rosen, M., M34, S45
Rosen, R., B5, S119
Rosenstein, I., S45
Ross, J., Q5
Rothschild, E., S49
Rowland, J. M., S180
Ruben, B. D., I70
Ruderman, H. D., C45, C87, S221, T12
Ruppenthal, K. M., S166
Rusche, P., M7
Ruth, J., A77
Ryterban, D., E66, E68

Sackson, S., A5, B22, E57, M140, T108, V3
Sampson, B., L50
Sampson, R. T., N8
Sanders, B., O7
Sarthory, J. A., S214
Schainker, S. A., N39
Schellenberger, R. E., M37, U6
Schild, E. O., G27, P10
Schonell, F. J., B38, J5, P4
Schorr, B., I3, O1
Schrieber, A. N., T77
Schwartz, H., R34
Scola, E., P18
Scott, G., C109
Scott, R., B5
Sears, D. W., N23

Seasholes, B., W37
Shapiro, M. J., R35
Shirts, G., P87, S169
Short, R. R., C73
Shukraft, R., A74, I5
Shure, G. H., T60
Silvern, L. C., S223
Simmons, J., S78
Sisson, R. L., I17, M49, M64, R39, T80
Smith, C. N., D20, P122
Smith, S., C148
Smith, W. N., I64
Smoker, P., I75
Soilec, J., M26
Solomon. E., F18, G47
Sommer, R., B37
Sord, B. H., D15
Spackman, A., S21
Spelder, B. E., A77
Springle, H. A., A49, A50, A51, A58, A59,
    B83, C62, C63, F31, F41, F42, F48, F70,
    F71, H34, H35, L5, O47, O48, O49, R45,
    R46, T63, T69, T97, T98, T116
Stapleton, R. N., M28
Steffens, H., C150
Stephens, J. A., B81
Stephenson, A., F20, M157
Stitelman, L., A34, A35, B63, B64, C112,
    C113, D12, P93
Stoane, J. S., F54
Summit, R. K., A16
Suransky, L., C66
Sutton, D. F., P99
Swanson, L. A., P35

Talbot, A., I70
Talbot, R. J., G58, G59, G60, G61
Tansey, P. J., H26, L46, S32, T21, W47
Tart, J., B37
Taylor, J. L., L52
Taylor, T., D56
Taylor, W., A42, H15
Thiagarajan, E68, E79
Thomas, A. E., B32, D40, D41, F66, O9,
    S152, S154, S219, V14
Thomas, O. G., A23, A24, D33
Thompson, H. E., C144
Thorelli, H. B., I68, I80
Tidswell, W. V., H19
Tillman, K. G., G9
Toll, D., G36
Tolley, H., T91

Totten, C., S187
Towle, J. W., W8
Townsend, K., R31
Trbovich., E79
Treuhertz, F., L19
Trilling, H., D44, S119
Tritten, D. E., F68, S84
Tusson, J. R., T96
Twelker, P. A., C100, M38, P77, T102

Unwin, D. J., U5

Vaccaro, M. J., G54
Vance, S., M23, M24
Van Der Eyken, W., B52
Vaughan, J. A., E79, U12
Vincent, W. H., S74
Vine, N. P., S199
Vines, E. F., I64

Wade, D. E., S214
Wagland, P. J., M135
Walford, R., B67, B71, D32, E87, N42, N43,
    R8, S65, W21
Walker, A. H., M33, P32, R34
Walker, J., S45
Walsh, J. M., R43
Warshaw, L. D., G57, G58, G60, G61
Washburn, J., A74, I5
Webb, O55
Weikmann, C., K7
Weiner, K., H18
Weir, L. C., C4
Wells, C., P11
Wells, H. G., L40
West, A. P., B2, C82, M42, S37
Weston, R. F., S80
Whaley, B., M137
Wheeler, O55
Whithed, M. H., D20, P122, W37
Whitton, H. J. G., A25
Whybark, D. C., S166
Whysall, P. T., C34, P47, O6
Wiggins, T. W., E27, E28
Williams, D. F., S206
Williams, W. J., E59, M54, P34, S212
Willingham, J. J., A3
Wilson, J. T., S218
Winters, P. R., C27, P110
Witkin, A., C20
Wohlking, W., H2, H3, H4
Wollaston, J. G. F., B84

Wood, R. C., W37
Wright, J., E9
Wynn, R., D13

Young, J., E24
Yount, D. E., D27, D54, M130, P7, S204

Zaltman, G., C119, C120
Zif, J. J., C126, I22, M33, M55, P32, P61,
 R34, S10, S208, Y4
Ziverneman, J., E9
Zocchi, L., B13, S48
Zoll, A. A., A52, A71, E56, E73, F14, G24,
 H36, I9, I10, I69, J7, M63, O54

5.4 Manufacturers and suppliers

ABBATT TOYS LTD,
94 Wimpole Street,
London W.1

ABT ASSOCIATES INC,
55 Wheeler Street,
Cambridge,
Massachusetts 02138, USA

ACADEMIC GAMES ASSOCIATES
INC,
430 East Thirty-third Street,
Baltimore,
Maryland 21218, USA

ACO GAMES DIVISION,
ALLEN CO INC
1208 East Market Street,
Charlottesville,
Virginia 22901, USA

ADDISON-WESLEY INC,
West End House,
11 Hills Place,
London W.1

ADMINISTRATIVE RESEARCH
ASSOCIATES,
Box 3,
Deerfield,
Illinois, USA

ADVANCED RESEARCH PROJECTS
AGENCY,
Defence Department,
Washington D.C., USA

AIR CANADA,
Place Ville Marie,
Montreal 2,
Canada

ALLYN AND BACON INC,
470 Atlantic Avenue,
Boston,
Massachusetts, USA

ALBION,
c/o D. Turnbull,
Flat 13, Gilmerton Court,
Trumpington Road,
Cambridge CB2 2HQ

AMERICAN ANTHROPOLOGICAL
ASSOCIATION,
Anthropological Curriculum Study Project,
5632 Kimbark Ave,
Chicago,
Illinois 60637, USA

AMERICAN BEHAVIORAL
SCIENTIST,
Sage Publications Inc,
275 South Beverly Drive,
Beverly Hills,
California 90212, USA

AMERICAN EDUCATION
PUBLICATIONS,
Xerox Education Group,
Education Center,
Columbus,
Ohio 43216, USA

AMERICAN INSTITUTES FOR
RESEARCH,
Box 1113,
Palo Alto,
California 94302, USA

AMERICAN INSTITUTE OF
PLANNERS,
917 15th Street NW,
Room 800,
Washington D.C. 20005, USA

AMERICAN MANAGEMENT
ASSOCIATION,
135 West 50th Street,
New York,
New York 10020, USA

AMERICAN MARKETING
ASSOCIATION
230 North Michigan Avenue,
Chicago 60601,
Illinois, USA

ANTIOCH BOOKPLATE CO,
Yellow Springs,
Ohio 45387, USA

APPLETON-CENTURY-CROFTS INC,
440 Park Avenue South,
New York,
New York 10016, USA

APPLIED SIMULATIONS
INTERNATIONAL
Suite 900,
110 17th Street North West,
Washington D.C. 20036, USA

ARIEL PRODUCTIONS LTD
St. Giles House,
Poland Street,
London W.1

ARIZONA STATE UNIVERSITY,
Temple,
Arizona 85281, USA

ARMY STRATEGY AND TACTICS
ANALYSIS GROUP,
The Pentagon,
Washington, USA

EDWARD ARNOLD LTD,
25 Hill Street,
London W1X 8LL

E. J. ARNOLD AND SON LTD,
Butterley Street,
Leeds LS10 1AX,
Yorkshire

ASSOCIATION FOR JEWISH
YOUTH,
33 Henriques Street,
Commercial Road,
London E.1

ASL PASTIMES,
Dudley,
Worcestershire

AURORA PLASTICS OF
CANADA LTD,
PO Box 243,
Rexdale,
Ontario, Canada

AUTOTELIC INSTRUCTIONAL
MATERIALS PUBLISHERS,
New Haven,
Connecticut, USA

AVALON-HILL INC,
4517 Harford Road,
Baltimore,
Maryland, USA

B.L. INDUSTRIAL TRAINING
AIDS LTD,
30 Fleet Street,
London E.C.4

BANK OF CALIFORNIA,
11th Floor,
400 California Street,
San Francisco,
California 94120, USA

HARVEY L. BARNETT,
1026 Starlite Lane,
Yuba City,
California 95991, USA

BATH YOUTH AND COMMUNITY
SERVICE,
9a York Street,
Bath, Somerset

BEHAVIORAL SCIENCES
LABORATORY,
College of Social & Behavioral Sciences,
Ohio State University,
404-B West 17th Avenue,
Columbus,
Ohio 43210, USA

BEHAVIORAL SIMULATION AND
GAMING GROUP,
Political Science Department,
Peoples Avenue Complex,
Building D,
Rensseleer Polytechnic Institute,
Troy,
New York 12181, USA

BELL BAXTER SENIOR HIGH
SCHOOL,
Cupar,
Fife

BERKSHIRE COLLEGE OF
EDUCATION,
Woodley Avenue,
Woodley,
Reading,
Berkshire

BINGLEY COLLEGE OF
EDUCATION,
Bingley,
Yorkshire

A AND C BLACK LTD,
4/5/6 Soho Square,
London W1V 6AD

BOARD OF COOPERATIVE
EDUCATIONAL SERVICES,
Northern Westchester County,
845 Fox Meadow Road,
Yorktown Heights,
New York 10598, USA

BOBBS-MERRILL COMPANY INC
4300 West 62nd Street,
Indianapolis,
Indiana 46268, USA

BOOZ, ALLEN AND HAMILTON,
Room 1700 Field Building,
135 South La Salle Street,
Chicago,
Illinois 60603, USA

MILTON BRADLEY,
Springfield,
Massachusetts 01101, USA

BRISTOL DISTRICT METHODISTS
ASSOCIATION OF YOUTH CLUBS,
Bristol,
Somerset

BRISTOL UNIVERSITY,
Bristol 8,
Somerset

BRISTISH BROADCASTING
CORPORATION,
Broadcasting House,
London W1A 1AA

BRITISH EUROPEAN AIRLINES,
Bealine House,
Ruislip,
Middlesex

BROOKLANDS COUNTY
TECHNICAL COLLEGE,
Heath Road,
Weybridge,
Surrey

BRUNEL UNIVERSITY,
Further Education Group,
Woodlands Avenue,
Acton,
London W.3

CADACO INC,
310 West Park Street,
Chicago,
Illinois 60607, USA

CAREERS RESEARCH AND
ADVISORY CENTRE,
Bateman Street,
Cambridge,
Cambridgeshire

CARNEGIE INSTITUTE OF
TECHNOLOGY,
Schenley Park,
Pittsburgh,
Pennsylvania 15213, USA

CARNEGIE-MELLON UNIVERSITY,
School of Industrial Administration,
Pittsburgh,
Pennsylvania 15213, USA

CASTELL BROS LTD,
15/17 St Cross Street,
Hatton Garden,
London E.C.1

CATHOLIC SCHOOL JOURNAL,
Pittsburgh,
Pennsylvania, USA

CEDAR RAPIDS COMMUNITY
SCHOOL DISTRICT,
346 2nd Avenue, South West,
Cedar Rapids,
Iowa 52404, USA

CENTER FOR ENVIRONMENTAL
QUALITY MANAGEMENT,
Cornell University,
302 Hollister,
Ithaca,
New York 14850, USA

CENTRE FOR STRUCTURAL
COMMUNICATION,
Kingston Polytechnic,
Kingston-on-Thames,
Surrey

CENTRAL MICHIGAN
EDUCATIONAL RESEARCH
COUNCIL,
Michigan, USA

CHAD VALLEY,
Barclay Toy Group Ltd,
Morden Road,
London SW19 3XL

CHICAGO PUBLIC SCHOOL
SYSTEM,
Chicago,
Illinois, USA

CITY UNIVERSITY,
St. John Street,
London E.C.1

COCA COLA EXPORT CORP,
Atlantic House,
7, Rockley Road,
London W.14

COLLIER-MACMILLAN PUBLISHERS,
35 Red Lion Square,
London WC1R 4SG

COLUMBIA UNIVERSITY,
Teachers College,
525 West 120th Street,
New York,
New York 10027, USA

COMMUNICATIONS WORKERS
OF AMERICA,
Education Director,
1925 K Street North West,
Washington D.C. 20006, USA

COMMUNITY MAKERS INC,
13 West 89th Street,
New York,
New York 10024, USA

COMMUNITY SERVICE
VOLUNTEERS,
Toynbee Hall,
28, Commercial Street,
London E1 6BR

COMPUTER CONGENERICS
CORPORATION,
4545 Lincoln Boulevard,
Oklahoma City,
Oklahoma 73105, USA

CONDOR TOYS LTD,
Wellington Road,
London Colney,
Hertfordshire

PROFESSOR J. CONWAY,
Gonville and Caius College,
Cambridge

J. R. COOPER,
7 St. Georges Avenue,
Rugby,
Warwickshire

CORNELL UNIVERSITY,
Ithaca,
New York 14850, USA

COURIER,
UNESCO, Place de Fontenoy,
Paris 7e, France

CREATIVE COMMUNICATIONS
AND RESEARCH,
460-35th Avenue,
San Francisco,
California 94121, USA

CREATIVE PLAYTHINGS,
Princeton,
New Jersey, USA

CREATIVE SPECIALITIES INC,
126 Jericho Turnpike,
Floral Park,
New York State 11001, USA

CREATIVE STUDIES INC,
167 Covey Road,
Boston,
Massachusetts 02146, USA

CUNA INTERNATIONAL INC,
Madison,
Wisconsin, USA

CURRICULUM DEVELOPMENT
ASSOCIATES INC,
Suite 414,
1211 Connecticut Avenue North West
Washington D.C. 20036, USA

CURRICULUM DEVELOPMENT
CENTER,
Wellesley School System,
Seawood Road,
Wellesley Hills,
Massachusetts, USA

DECALSET,
16 Davenport Road,
Sidcup,
Kent

DEFENCE OPERATIONAL
ANALYSIS ESTABLISHMENT,
Byfleet,
Surrey

DENYS FISHER TOYS LTD,
Thorpe Arch Trading Estate,
Boston Spa,
Yorkshire

DEPARTMENT OF THE
ENVIRONMENT,
Central Training Branch,
Neville House,
Page Street,
London

DERBYSHIRE YOUTH SERVICE,
Matlock,
Derbyshire

DIDACTIC SYSTEMS INC,
P.O. Box 500,
Westbury,
Long Island,
New York 11590, USA

DOUBLEDAY & CO INC,
501 Franklin Avenue,
Garden City,
New York,
New York, USA

DOUGLAS AIRCRAFT
CORPORATION,
3855 Lakewood Boulevard,
Long Beach,
California 90801, USA

DOW JONES, IRWIN,
1818 Ridge Road,
Houseword,
Illinois, USA

DREXEL INSTITUTE OF
TECHNOLOGY,
Philadelphia,
Pennsylvania 19103, USA

DUNCHURCH INDUSTRIAL STAFF
COLLEGE,
Dunchurch,
Rugby,
Warwickshire

DUQUESNE UNIVERSITY,
600 Forbes Avenue,
Pittsburgh,
Pennsylvania 15219, USA

DYNAMIC DESIGN INDUSTRIES,
1433 North Central Park
Anaheim,
California, USA

DYNASTY INTERNATIONAL INC,
815 Park Avenue,
New York,
New York 10021, USA

EARLHAM COLLEGE,
Richmond,
Indiana 47374, USA

ED/COM SYSTEMS INC,
145, Witherspoon Street,
Princeton,
New Jersey 08540, USA

EDINBURGH UNIVERSITY,
29 George Square,
Edinburgh 8

EDITIONS OUVRIERES,
12 Avenue Soeur-Rosalie,
Paris 13, France

EDUCATION AND TRAINING
CONSULTANTS CO,
P.O. Box 49899,
Los Angeles,
California 90049, USA

EDUCATION DEVELOPMENT
CENTER INC,
15 Mifflin Place,
Cambridge,
Massachusetts 02138, USA

EDUCATIONAL GAMES CO,
P.O. Box 363,
Peekskill,
New York 10566, USA

EDUCATIONAL INNOVATIONS,
4 Harrogate Court,
Droitwich Close,
Sydenham Hill,
London SE26 STL

EDUCATIONAL METHODS INC,
20 East Huron Street,
Chicago,
Illinois 60611, USA

EDUCATIONAL RESEARCH,
Newnes Educational Publishing Co,
Hamlyn House,
42 The Centre,
Feltham, Middlesex

EDUCATIONAL SERVICES INC,
15 Mifflin Place,
Cambridge,
Massachusetts, USA

EDUCATIONAL SIMULATION
LABORATORY,
College of Education,
Ohio State University,
Columbus,
Ohio, USA

EDUCATIONAL SUPPLY
ASSOCIATION LTD,
The Pinnacles,
Harlow, Essex

El CAPITAN HIGH SCHOOL,
Lakeside,
California, USA

C. ELGOOD,
21, Cork Street,
London W1X 1HB

ENGLISH UNIVERSITIES PRESS
St. Pauls' House,
Warwick Lane,
London E.C.4

ENTELEK INC,
42, Pleasant Street,
Newburyport,
Massachusetts 01950, USA

ENVIRONMENTAL SIMULATION
LABORATORY,
611 Church Street,
Ann Arbor,
Michigan 48104, USA

ENVIROMETICS SCIENCE
RESOURCES CORPORATION,
1100 17th Street North West,
Washington D.C. 20036, USA

ESSO PETROLEUM COMPANY LTD,
Education Services,
Victoria Street,
London S.W.1

EUROPEAN RESEARCH GROUP
ON MANAGEMENT,
53 rue de la concorde,
Brussels 5, Belgium

THE E.V.R. PARTNERSHIP,
1 Hanover Square,
London W.1

EVANS BROS LTD,
Montague House,
Russell Square,
London W.C.1

HOWARD FARROW
CONSTRUCTION LTD,
Highfield Road,
London N.W.11

FEARON PUBLISHERS,
6 Davis Drive,
Belmont,
California 94002, USA

FINANCIAL TIMES,
Bracken House,
10, Canon Street,
London E.C.4

FOREIGN POLICY ASSOCIATION,
345 East 46th Street,
New York,
New York 10017, USA

THE FREE PRESS OF GLENCOE,
60 Fifth Avenue,
New York,
New York 10011, USA

GALT TOYS LTD,
Cheadle, Cheshire

GAMES AND PUZZLES,
Edu-Games (U.K.) Ltd,
P.O. Box 4,
London N6 4DF

GAMES CENTRAL,
Abt Associates Inc,
55, Wheeler Street,
Cambridge,
Massachusetts, USA

GAMES RESEARCH INC,
48 Wareham Street,
Boston 18,
Massachusetts, USA

GAMESCIENCE
No longer trading

GARRARD PUBLISHING CO,
Champaign,
Illinois 61820, USA

GENERAL ELECTRIC CO,
7735 Old Georgetown Road,
Bethesda,
Maryland, USA

GEOGRAPHICAL ASSOCIATION,
343 Fulwood Road,
Sheffield 10

G. I. GIBBS,
5 Errington,
Moreton-in-Marsh,
Gloucestershire

H. P. GIBSON,
319 Kingston Road,
London

GLAMORGAN YOUTH SERVICE,
County Hall,
Cathay's Park,
Cardiff

MARY GLASGOW PUBLICATIONS
LTD,
140 Kensington Church Street,
London W8 4BN

GLASGOW UNIVERSITY,
4 University Gardens,
Glasgow W.2

GREATERMANS STORES LTD,
P.O. Box 5460,
Johannesburg, South Africa

GRIFFIN AND GEORGE,
Frederick Street,
Birmingham 1,
Warwickshire

GUARDIAN BUSINESS SERVICES,
192 Gray's Inn Road,
London W.C.1

GUIDON GAMES,
P.O. Box 1123,
Evansville, Indiana 47713, USA

HABITAT,
Hithercroft Road,
Wallingford,
Oxfordshire

HANAN & SON,
P.O. Box 1234,
Grand Central Station,
New York,
New York 10017, USA

HARCOURT, BRACE,
JOVANOVICH INC,
757 Third Avenue,
New York,
New York 10017, USA

HARLECH TELEVISION,
Television Centre,
Cathedral Road,
Cardiff

RUPERT HART-DAVIS LTD,
Park Street,
St Albans,
Hertfordshire

HARVARD BUSINESS REVIEW
Soldier's Field,
Boston,
Massachusetts 02163, USA

HARVARD UNIVERSITY,
220 Alewife Brook Parkway,
Cambridge,
Massachusetts 02138, USA

C. HARVEY,
30 Goreway Road,
Walsall,
Staffordshire

HARWELL ASSOCIATES,
Covent Station,
New Jersey 07961, USA

HATFIELD POLYTECHNIC,
Hatfield,
Hertfordshire

M. HAYES,
85, Barncliffe Cresent,
Sheffield S10 4DB,
Yorkshire

HEAL'S,
196 Tottenham Court Road,
London W.1

D. C. HEATH & CO INC,
125 Spring Street,
Lexington,
Massachusetts 02173, USA

PIET HEIN,
Games manufactured by:
Skjode,
Skjearn, Denmark

C. HEMMING,
20 Hilltop Court,
Wilmslow Road,
Fallowfield,
Manchester M14 6LH

HERDER AND HERDER INC,
1221 Avenue of the Americas,
New York,
New York 10019, USA

HEYDEN AND SON LTD,
Spectrum House,
Alderton Crescent,
London N.W.4

HIGH SCHOOL GEOGRAPHY
PROJECT,
P.O. Box 1095,
Boulder,
Colorado 80302, USA

HOLMES McDOUGALL LTD,
30, Royal Terrace,
Edinburgh, USA

HOLT, RINEHART & WINSTON
INC,
383 Madison Avenue,
New York,
New York 10017, USA

HONEYWELL INFORMATION
SYSTEMS LTD,
Honeywell House,
Great West Road,
Brentford,
Middlesex

HORNSEY COLLEGE OF ART,
Crough End Hill,
London N.8

HORTON WATKINS HIGH SCHOOL,
La Due,
Missouri, USA

HOUGHTON MIFFLIN CO,
Educational Division,
110 Tremont Street,
Boston,
Massachusetts 02107, USA

HOUSE OF STYLE,
13 Hailey Road,
Erith,
Kent

IDEAS IN GEOGRAPHY,
Department of Geography,
University of Nottingham,
Nottingham

IMPERIAL COLLEGE OF SCIENCE
AND TECHNOLOGY
University of London,
Prince Consort Road,
London S.W.7

IMPERIAL OIL AND GREASE
COMPANY INC,
6505 Wilshire Boulevard,
Los Angeles,
California 90048, USA

INDIANA UNIVERSITY,
Bloomington,
Indiana 47401, USA

INDUSTRIAL COLLEGE OF THE
ARMED FORCES,
Fort McNair,
Washington D.C., USA

INDUSTRIAL AND COMMERCIAL
TRAINING,
The Old Sun,
Guilsborough,
Northampton NN6 8PY

INFINITY COMMUNICATIONS LTD,
34 Sussex House,
Charlton Street,
London N.W.1

INFORMATION RESOURCES INC,
1675 Massachusetts Avenue,
Cambridge,
Massachusetts 02138, USA

INSTAD LTD,
9650 Midtown Plaza Station,
Rochester,
New York 14604, USA

INSTITUTE OF CHARTERED
ACCOUNTANTS,
56 Goswell Road,
London E.C.1

INSTITUTE OF DEVELOPMENT
STUDIES,
University of Sussex,
Falmer,
Brighton

INSTITUTE OF PRODUCTION
ENGINEERS,
10 Chesterfield Street,
London W.1

INSTRUCTIONAL SIMULATIONS INC,
2147 University Avenue,
St. Paul,
Minnesota 55114, USA

INTELLECT (UK) LTD,
49 Great Marlborough Street,
London W1V 1DB

INTERACT,
P.O. Box 262,
Lakeside,
California 92040, USA

INTER-ACTION,
156 Maiden Road,
London NW5

INTERNATIONAL ACADEMIC
GAMES,
440 Las Olas Boulevard,
Fort Lauderdale,
Florida 33301, USA

INTERNATIONAL BUSINESS
MACHINES INC,
White Plains,
New York, USA

IMPERIAL CHEMICAL
INDUSTRIES LTD,
Albert Embankment,
London

INTERNATIONAL COMPUTERS LTD,
Bridge House,
Putney,
London

INTERNATIONAL LEARNING
CORPORATION,
245 South West 32nd Street,
Fort Lauderdale,
Florida 33301, USA

INTERNATIONAL TEXTBOOK CO,
Scranton,
Pennsylvania 18515, USA

INTERTEXT GROUP LTD,
Intertext House,
Stewarts Road,
London S.W.8

INVICTA PLASTICS,
Oadby,
Leicester

IPC BUSINESS AND INDUSTRIAL
TRAINING,
161 Fleet Street,
London E.C.4

R. D. IRWIN INC,
1818 Ridge Road,
Homewood,
Illinois, USA

I.W.A. RESCHENSCHEIBERFABRIK,
F. Riehle KG73,
Esslingen, Germany

JACKDAW PUBLICATIONS LTD,
30 Bedford Square,
London WC1B 3EL

L. JACOBS,
2500 Campus Road,
Honolulu,
Hawaii 96822, USA

J. JACQUES AND SON LTD,
Whiteheather Works,
Whitehouse Road,
Thornton Heath,
Surrey

G. JEFFREY,
15 Rusholme Road,
London S.W.15

THE JEWISH AGENCY,
515 Park Avenue,
New York, USA

J. K. JONES,
10, Kildare Court,
Kildare Terrace,
London W2 5JU

JOHNS HOPKINS UNIVERSITY,
3505 North Charles Street,
Baltimore,
Maryland 21212, USA

JOINT COUNCIL ON ECONOMIC
EDUCATION,
1212 Avenue of the Americas,
New York,
New York 10036, USA

JOURNEY GAMES,
35, Laburnam Road,
Maidenhead,
Berkshire

JUMBO GAMES,
Hausemann and Hotte,
Kromboomffloot 53-61,
Amsterdam, Holland

KANSAS STATE TEACHERS
COLLEGE,
Emporia,
Kansas 66801, USA

KDI INSTRUCTIONAL SYSTEMS
INC,
3077 Silver Drive,
Columbus,
Ohio 43224, USA

KMS INDUSTRIES INC,
220 East Huron Street,
Ann Arbor,
Michigan 48108, USA

JOHN LAING & SON LTD,
Page Street,
London N.W.7

LEARNING INSTITUTE OF
NORTH CAROLINA,
Duke University,
Durham,
North Carolina, USA

A. D. LITTLE INC,
Cambridge,
Massachusetts, USA

LITTLE, BROWN, & CO
34 Beacon Street,
Boston,
Massachusetts, USA

LIVERPOOL EDUCATIONAL
PRIORITY AREA PROJECT,
Paddington Comprehensive School,
Liverpool L7 3EA

LOCKHEED EDUCATION SYSTEMS,
Organisation 56-70,
P.O. Box 504,
Sunnyvale,
California 94088, USA

LOCKHEED PALO ALTO
RESEARCH LABORATORY,
3251 Hanover Street,
Palo Alto,
California, USA

LONGMAN GROUP LTD,
Burnt Mill,
Harlow, Essex

R. LOOMIS,
8149 East Thomas Road,
Scottsdale,
Arizona 85251, USA

R. LUNSTEDT,
1750 South White Road,
San Jose,
California, USA

LYONS AND CARNAHAN,
407 East 25th Street,
Chicago,
Illinois 60616, USA

MACALESTER COLLEGE,
Dept. Political Science (Simulation Center),
St. Paul,
Minnesota, USA

MACDONALD & CO LTD,
BPC Publishing Ltd,
St. Giles House,
49-50 Poland Street,
London W1E 6JZ

THE MACMILLAN CO,
866 Third Avenue,
New York,
New York 10022, USA

MACOS,
Educational Development Center,
15 Mifflin Place,
Cambridge,
Massachusetts, 02138, USA

MACSCO,
P.O. Box 382,
Locust Valley,
New York 11560, USA

MAGNAVOX ELECTRONICS
CO LTD,
Magnavox House,
Alfred's Way,
Barking,
Essex

MANAGEMENT ASSOCIATES,
4 Fitzalan Place,
Cardiff,
Glamorgan

MANAGEMENT DEVELOPMENT
INC,
148 East Lancaster Avenue,
Wayne,
Pennsylvania 19087, USA

MANAGEMENT GAMES LTD,
38, Bedford Place,
London W.C.1

MANCHESTER COLLEGE OF
EDUCATION,
Long Millgate,
Manchester M3 1SD

MARCT COMPANY,
1111 Maple Avenue,
Turtle Creek,
Pennsylvania, USA

MARIA GREY COLLEGE OF
EDUCATION,
Twickenham,
Middlesex

MARI-RAMA INC,
2859 Bird Avenue Suite 5,
Coconut Grove,
Florida 33133, USA

MARTIN-MARIETIA CORP;
Mail Point 211,
P.O. Box 5837,
Orlando,
Florida 32806, USA

MASSACHUSETTS INSTITUTE OF
TECHNOLOGY,
Cambridge,
Massachusetts, USA

MASSACHUSETTS INSTITUTE OF
TECHNOLOGY PRESS,
Cambridge,
Massachusetts, USA

MATCHBOX TOYS LTD,
Lesney Products & Co Ltd,
Lee Conservancy Road,
London E.9

MATHEMATICAL PIE,
100 Burman Road,
Shirley,
Solihull,
Warwickshire

MATHEMATICS IN SCHOOL,
The Longman Group,
5 Bentinck Street,
London W.1

MATHEMATICS TEACHING,
Vine Street Chambers,
Vine Street,
Nelson,
Lancashire

MATHS LEARNING SYSTEMS,
10, Sundays Hill,
Lower Almondsbury,
Bristol BS12 4DR,
Somerset

McGRAW-HILL LTD,
Shoppenhangers Road,
Maidenhead,
Berkshire

MERIT TOYS,
J & L Randall Ltd,
Cranborne Road,
Potters Bar,
Hertfordshire

CHARLES E. MERRILL INC,
1300 Alum Creek Drive,
Columbus,
Ohio 43216, USA

MERSEYSIDE TRAINING CENTRE,
Spekehall Road,
Speke,
Liverpool 24

METHUEN & CO LTD,
11 New Fetter Lane,
London EC4P 4EE

METROPOLITAN LIFE INSURANCE
CO,
1 Madison Avenue,
New York,
New York 10010, USA

MICHIGAN STATE UNIVERSITY,
East Lancing,
Michigan 48823, USA

S. J. MILLER CO INC,
P.O. Box 130,
Coney Island Station,
Brooklyn,
New York 11224, USA

MINIATURE WARFARE,
36 Kennington Road,
Lambeth Walk,
London S.E.

MINISTRY OF PUBLIC BUILDING
AND WORKS,
Central Training Branch,
Neville House,
Page Street,
London S.W.1

MINISTRY OF TRANSPORT
(now part of),
Department of the Environment,
2 Marsham Street,
London W.1

MOLOR PRODUCTS CORPORATION,
P.O. Box 709,
Glen Ellyn,
Illinois 70137, USA

MOSESSON GAMES LTD,
Creeting Road,
Stowmarket,
Suffolk

MOTHER,
Odhams Press Ltd,
189 High Holborn,
London W.C.1

MOUNT PLEASANT HIGH SCHOOL,
1750 Sal White Road,
San Jose,
California, USA

NATIONAL ACADEMIC GAMES
PROJECT,
Nova University,
South West College Avenue,
Fort Lauderdale,
Florida 33314, USA

NATIONAL ASSOCIATION OF
BOYS CLUBS,
17 Bedford Square,
London W.C.1

NATIONAL ASSOCIATION OF
YOUTH CLUBS,
30 Devonshire Street,
London W.1

NATIONAL COMPUTING
CENTRE LTD,
Quay House,
Quay Street,
Manchester M3 3HU

NATIONAL 4-H CENTER,
7100 Connecticut Avenue,
Washington D.C. 20015, USA

NATIONAL PORTS COUNCIL,
1 New Oxford Street,
London W.1

NATIONWIDE INSURANCE,
246 North High Street,
Columbus,
Ohio 43215, USA

T. NELSON AND SONS LTD,
36 Park Street,
London W1Y 4DE

NEW CENTURY INC,
440 Park Avenue,
New York,
New York 10016, USA

NEW MEXICO STATE UNIVERSITY,
Albuquerque,
New Mexico, USA

NEW SOCIETY,
New Science Publications Ltd,
Cromwell House,
Fulwood Place,
High Holborn,
London W.C.1

NEW TOWN,
c/o B. R. Lawson,
57 Cherry Hill Road,
Norwich,
Connecticut 06360, USA

NEW UNIVERSITY OF ULSTER,
Coleraine,
Northern Ireland

NEW YORK STATE EDUCATION
DEPARTMENT,
Albany,
New York 12224, USA

NORTH IDAHO JUNIOR COLLEGE,
1000 West Gordon Avenue,
Coeur d'Alene,
Idaho 83814, USA

NORTHWESTERN UNIVERSITY,
1834 Sheridan Road,
Evanston,
Illinois 60201, USA

NOVA LEARNING CORPORATION,
Nova,
Florida, USA

NOVA HIGH SCHOOL
ACADEMIC GAMES DIRECTOR,
3600 Southwest 70th Avenue,
Fort Lauderdale,
Florida 33314, USA

NOVA SCOTIA TECHNICAL
COLLEGE,
Nova Scotia, Canada

OHIO STATE UNIVERSITY,
Columbus,
Ohio, USA

O.K.H. (West Germany)
No longer exists

OKLAHOMA STATE UNIVERSITY,
Stillwater,
Oklahoma, USA

OLCOTT FORWARD INC,
234 North Central Avenue,
Hartsdale,
New York 10530, USA

OLIVER AND BOYD LTD,
23 Ravelston Terrace,
Edinburgh EH4 3TJ

D. R. OLSON,
252 Bloor Street West,
Toronto 5, Canada

ORCHARD TOYS LTD,
The Old Stables,
Main Street,
Keyworth,
Nottinghamshire

OREGON STATE UNIVERSITY,
Corvallis,
Oregon 97331, USA

OXFAM,
274 Banbury Road,
Oxford OX2 7DZ

OXFORD UNIVERSITY PRESS,
37 Dover Street,
London W.1

PÄDAGOGISCHE BEITRÄGE,
Georg Westermann Verlag,
33 Braunschweig,
Georg-Westermann-Allee 66,
Postfachs 7049, West Germany

PALITOY LTD,
Coalville,
Leicester

PANTHER BOOKS LTD,
Park Street,
St. Albans,
Hertfordshire

PANZERFAUST,
c/o D. Lowry,
P.O. Box 1123,
Evansville,
Indiana 47713, USA

PARKER BROS INC,
190 Bridge Street,
Salem,
Massachusetts 01970, USA

H. PATTERSON,
7 Cambridge Road,
Beaconsfield,
Buckinghamshire

PEACE RESEARCH CENTRE,
58 Parkway,
London N.W.1

PENGUIN BOOKS LTD,
Harmondsworth,
Middlesex

PENNSYLVANIA STATE
UNIVERSITY,
211 Boucke Building,
University Park,
Pennsylvania,
Pennsylvania 6820, USA

PEPYS GAMES,
Castell Brothers Ltd,
15 Saint Cross Street,
London E.C.1

PERSONNEL & GUIDANCE
JOURNAL,
607 New Hampshire Avenue, North West,
Washington,
Washington D.C. 20009, USA

PETER PAN PLAYTHINGS LTD,
4-10 Rodney Street,
London N.1

PHILMAR LTD,
Beachy Road,
Old Ford,
London E3 2NX

J. PIGGOTT,
Jesus College,
Cambridge

PLAN B CORPORATION,
3090 Coney Island Avenue,
Brooklyn,
New York, USA

PLANNING AND TRANSPORT
RESEARCH AND COMPUTATION
LTD,
40 Grosvenor Gardens,
London

PLEASANT HILL HIGH SCHOOL,
3100 Oak Park Boulevard,
Pleasant High,
California 94573, USA

POLYCON BUILDING INDUSTRY
CONSULTANTS,
Triumph House,
189 Regent Street,
London W.1

PORTADOWN COLLEGE,
Portadown,
County Armagh,
Northern Ireland

JAMIE PORTER LTD,
Leeds, Yorkshire

RONALD PRESS CO,
79 Madison Avenue,
New York,
New York, USA

PRENTICE-HALL INC,
Englewood Cliffs,
New Jersey, USA

PRICE WATERHOUSE & CO,
3 Fredericks Place,
London E.C.2

PRINCETON UNIVERSITY,
Princeton,
New Jersey, USA

PROCTER AND GAMBLE LTD,
St. Nicholas Avenue,
Newcastle-on-Tyne, 3,
Co. Durham

PROGRAMMED INSTRUCTION,
Center for Programmed Instruction,
Teacher's College,
Columbia University,
New York,
New York 10027, USA

PROJECT SIMILE,
1150 Silverado Road,
La Jolla,
California 92037, USA

PSYCHOLOGY TODAY GAMES,
P.O. Box 4762,
Clinton,
Iowa 52732, USA

PUBLIC OPINION QUARTERLY,
Room 510 Journalism Building,
Columbia University,
New York,
New York 10027, USA

RAND CORPORATION,
1700 Main Street,
Santa Monica,
California 90406, USA

RANDA INC,
19 Melony Avenue,
Plain View,
New York 11803, USA

J & L RANDALL LTD,
Merit House, Cranborne Road,
Potters Bar,
Hertfordshire

RANDOM HOUSE,
201 East 50th Street,
New York 10022, USA

RAYTHEON COMPANY,
141 Spring Street,
Lexington,
Massachusetts 02173, USA

RCA CORPORATION,
30 Rockefeller Plaza,
New York,
New York 10020, USA

READER'S DIGEST,
Social Studies Editor,
Reader's Digest Educational Division,
Pleasantville,
New York 10570, USA

REAL WORLD LEARNING INC,
134 Sunnydale Avenue,
San Carlos,
California 94070, USA

J. REESE,
3235 West 17th Avenue,
Eugene,
Oregon 97402, USA

REMINGTON RAND DIVISION,
Sperry Rand Corporation,
P.O. Box 1000,
Blue Bell,
Pennsylvania 19422, USA

RENSSELAER POLYTECHNIC
INSTITUTE,
Troy,
New York, USA

RESEARCH ANALYSIS
CORPORATION,
McLean,
Virginia 22101, USA

RESEARCH TRIANGLE INSTITUTE,
Research Triangle Park,
North Carolina, USA

J. M. ROWLAND,
1545, Harmony Road,
Akron,
Ohio 44313, USA

ROYAL AIR FORCE
SCHOOL OF EDUCATION,
Upwood,
Huntingdonshire

ROYAL SOCIETY FOR THE
PROTECTION OF BIRDS,
The Lodge,
Sandy,
Bedfordshire

RUTGERS UNIVERSITY,
New Brunswick,
New Jersey 08903, USA

S.D.C.,
2500 Colorado Avenue,
Santa Monica,
California 90406, USA

S.G.S. ASSOCIATES,
8 New Row,
London

SAGE PUBLICATIONS INC,
275 South Beverly Drive,
Beverly Hills,
California 90212, USA

ST. MARYS & ST. PAULS
GEOGRAPHICAL SOCIETY,
St. Pauls College of Education,
Cheltenham,
Gloucestershire

SAN FRANCISCO THEOLOGICAL
SEMINARY,
San Anselmo,
California 94960, USA

SATLA PRODUCTS,
11 Honeyway,
Royston,
Hertfordshire

G. SCHIRMER,
4 East 49th Street,
New York,
New York 10017, USA

THE SCHOOL GOVERNMENT
PUBLISHING CO LTD,
Department B,
Merstham, Surrey

SCHOOLS COUNCIL,
160 Great Portland Street,
London W1N 6LL

SCIENCE RESEARCH ASSOCIATES,
259 East Erie Street,
Chicago,
Illinois 60611, USA

SCOTT, FORESMAN & CO,
Glenview,
Illinois 60025, USA

SCOTTISH YOUTH & COMMUNITY
SERVICE INFORMATION CENTRE,
67, York Place,
Edinburgh 1

SECKER AND WARBURG LTD,
14 Carlisle Street,
London W.1

SELCHOW & RIGHTER CO,
2215 Union Boulevard,
Bay Shore,
New York 11206, USA

SELLOTAPE LTD,
54/58 High Street,
Edgware,
Middlesex

SENIOR SECRETARIES LTD,
173 New Bond Street,
London W.1

SEVEN TOWNS LTD,
Axtell House,
23 Warwick Street,
London W.1

SHELTER,
86 Strand,
London WC2R OEQ

R. R. SHORT,
Whitworth College,
Spokane,
Washington 99218, USA

SIMILE II,
1150 Silverado Road,
La Jolla,
California 92037, USA

SIMPLAY PRODUCTS,
Norprint Ltd,
Boston,
Lincolnshire

SIMULATED ENVIRONMENTS INC.
3401 Market Street,
Philadelphia,
Pennsylvania 19104, USA

SIMULATION COUNCILS INC,
P.O. Box 1023,
La Jolla,
California 92037, USA

SIMULATION INC,
P.O. Box 140,
Carmel,
Indiana 46032, USA

SIMULATION SYSTEMS PROGRAM,
Teaching Research Division,
Oregon State System of Higher Education,
Monmouth,
Oregon 97361, USA

SIMULATIONS PUBLICATIONS
INC,
44 East 23rd Street,
New York,
New York 11010, USA

SIMULATIONS PUBLICATIONS
INC,
7, Alexander Drive,
Timperley,
Cheshire

SIMULMATICS CORPORATION INC,
16 East 41st Street,
New York 17,
New York, USA

SINAUER ASSOCIATES,
Stamford,
Connecticut, USA

SLOUGH COLLEGE OF
TECHNOLOGY,
William Street,
Slough,
Buckinghamshire

SMALL BUSINESS ADMINISTRATION
Washington D.C. 20025, USA

W. H. SMITH & SONS LTD,
Strand House,
Portugal Street,
London W.C.2

SOCIAL EDUCATION,
National Council for Social Studies,
1201 16th Street North West,
Washington D.C. 20036, USA

SOCIAL STUDIES CURRICULUM
STUDY CENTER,
1809 Chicago Avenue,
Evanston,
Illinois 60201, USA

SOCIETY FOR ACADEMIC
GAMING AND SIMULATION
IN EDUCATION AND TRAINING,
5, Errington,
Moreton-in-Marsh,
Gloucestershire

SOCIETY OF ACTUARIES,
208 South La Salle Street,
Chicago,
Illinois 60604, USA

SOFTWARE SCIENCES HOLDING
LTD,
17 Curzon Street,
London W.1

SOUTHAMPTON ADULT
EDUCATION & YOUTH
SERVICE,
Southampton,
Hampshire

SOUTHGATE TECHNICAL
COLLEGE,
High Street,
Southgate,
London N.14

SOUTHWESTERN PUBLISHING CO,
5101 Madison Road,
Cincinnati,
Ohio 45227, USA

SPACESHIP EARTH EXPLORATION,
P.O. Box 909,
Carbondale,
Illinois 62901, USA

J. W. SPEAR & SONS LTD,
The Sales Manager,
Green Street,
Enfield,
Middlesex

SPENCER AND HALSTEAD LTD,
Bridge Works,
Ossett,
Leeds,
Yorkshire

SPERRY-RAND CORPORATION,
1700 Main Street,
Santa Monica,
California 90406, USA

SPIRING ENTERPRISES LTD,
North Holmwood,
Dorking,
Surrey

SPORTS ILLUSTRATED GAMES,
Time-Life Building,
Rockefeller Center,
New York,
New York 10020, USA

STANFORD UNIVERSITY,
Stanford,
California 94305, USA

STATE EDUCATION DEPARTMENT,
Albany,
New York 12224, USA

STAUDI LTD,
61, Conway Road,
Taplow,
Buckinghamshire

STOCK EXCHANGE,
Public Relations Dept,
London EC2N 1HP

STRATEGY AND TACTICS,
7 Alexander Drive,
Timperley,
Altrincham,
Cheshire WA15 6NF

STUDY-CRAFT EDUCATIONAL
PRODUCTS,
R1-683,
Tusson Research Center,
Belle Chasse,
Louisiania 70037, USA

SUBBUTEO,
Dept 192 Langton Green,
Tunbridge Wells,
Kent

SUSSEX YOUTH OFFICES,
Wellington Square,
Hastings,
Sussex

SYRACUSE UNIVERSITY,
Syracuse,
New York 13210, USA

SYSTEM DEVELOPMENT
CORPORATION,
2500 Colorado Avenue,
Santa Monica,
California 90406, USA

SYSTEMS GAMING ASSOCIATES,
Lancing Apartments,
Ithaca,
New York 14850, USA

SYSTEMS RESEARCH LTD,
2 Richmond Hill,
Richmond,
Middlesex

TAF SPORTS GAMES LTD,
P.O. Box 106,
Guernsey,
Channel Islands

TEACHING RESEARCH DIVISION
OREGON STATE SYSTEM OF
HIGHER EDUCATION,
Monmouth,
Oregon 97361, USA

TEMPLE UNIVERSITY,
Philadelphia,
Pennsylvania 19122, USA

TEXAS TECHNICAL UNIVERSITY,
Lubbock,
Texas 79409, USA

THINK (GAMES) LTD,
Dept. 1,
1A Gautrey Road,
London S.E.15

THIRD MILLENNIA INC,
465 Woodland Heights,
Philadelphia,
Mississippi 39356, USA

THREE, FOUR, FIVE, LTD,
92a Old Street,
London EC1V 9AY

3M COMPANY,
St. Paul,
Minnesota 55101, USA

TIMES EDUCATIONAL
SUPPLEMENT,
Printing House Square,
London E.C.4

TOKYO CENTER FOR ECONOMIC
RESEARCH,
Tokyo,
Japan

TRAINING AND DEVELOPMENT
CENTER,
Two Pennsylvania Plaza,
New York,
New York 10001, USA

TRAVELERS INSURANCE
COMPANY,
One Tower Square,
Hartford,
Connecticut, USA

TRENTON STATE COLLEGE,
Trenton,
New Jersey 08625, USA

TRIANG LTD,
Havant,
Hampshire

TUBE INVESTMENTS LTD.
Personnel Department,
Woodbourne Grange,
21 Woodbourne Road,
Birmingham B17 8BZ

UNITARIAN UNIVERSALIST
ASSOCIATION,
25 Beacon Street,
Boston,
Massachusetts 02108, USA

UNITED STATES ARMY
LOGISTICS MANAGEMENT
CENTER,
Fort Lee,
Virginia 23891, USA

UNITED STATES DEFENCE
DEPARTMENT,
Washington D.C., USA

UNITED STATES DEPARTMENT
OF AGRICULTURE, FOREST
SERVICE,
Division of Fire Control,
Washington D.C. 20250, USA

UNITED STATES MARINE CORPS,
Navy Department,
Washington 20380, USA

UNITED STATES OFFICE OF
EDUCATION,
U.S. Dept. of Health, Education and
Welfare,
Washington,
Washington D.C. 20402, USA

UNITED STATES NAVAL
INSTITUTE,
Annapolis
Maryland, USA

UNITED STATES NAVAL WAR
COLLEGE,
U.S. Naval Base,
Newport,
Rhode Island, USA

UNITED STATES TRUST CO,
45 Wall Street,
New York,
New York, USA

UNIVERSAL PUBLICATIONS LTD,
581 Lordship Lane,
Wood Green,
London N.22

UNIVERSITY ASSOCIATES PRESS,
P.O. Box 615,
Iowa City,
Iowa 52240, USA

UNIVERSITY COUNCIL FOR
EDUCATIONAL ADMINISTRATION,
65 South Oval Drive,
Columbus,
Ohio, USA

UNIVERSITY OF ALBERTA,
Edmonton,
Alberta, Canada

UNIVERSITY OF ARIZONA,
Tucson,
Arizona, USA

UNIVERSITY OF BIRMINGHAM,
P.O. Box 363,
Birmingham B15 2T1,
Warwickshire

UNIVERSITY OF BRISTOL,
21 Berkeley Square,
Bristol 8,
Somerset

UNIVERSITY OF BRITISH
COLUMBIA,
British Columbia, Canada

UNIVERSITY OF CALGARY,
Room 517,
Dept. Curriculum and Instruction,
Faculty of Education,
Calgary 44,
Alberta, Canada

UNIVERSITY OF CALIFORNIA,
Santa Barbara,
California 93106, USA

UNIVERSITY OF CHICAGO,
1225 East 60th Street,
Chicago,
Illinois 60637, USA

UNIVERSITY OF CONNECTICUT,
Storrs,
Connecticut, USA

UNIVERSITY OF GEORGIA,
Athens,
Georgia 30601, USA

UNIVERSITY OF HAWAII,
Honolulu,
Hawaii, USA

UNIVERSITY OF ILLINOIS,
Urbana,
Illinois 61801, USA

UNIVERSITY OF LANCASTER,
Bailrigg,
Lancaster,
Lancashire

UNIVERSITY OF LETHBRIDGE,
Lethbridge,
Alberta, Canada

UNIVERSITY OF LIVERPOOL,
P.O. Box 147,
Liverpool L69 3BX

UNIVERSITY OF LONDON,
Senate House,
London W.C.1

UNIVERSITY OF MANCHESTER,
P.O. Box 88,
Sackville Street,
Manchester M60 1QD,
Lancashire

UNIVERSITY OF MASSACHUSETTS,
Amherst,
Massachusetts, USA

UNIVERSITY OF MIAMI,
Coral Gables,
Florida, USA

UNIVERSITY OF MICHIGAN,
Ann Arbor,
Michigan 48104, USA

UNIVERSITY OF NEBRASKA,
Lincoln,
Nebraska 65508, USA

UNIVERSITY OF NEVADA,
Reno,
Nevada, USA

UNIVERSITY OF NEWFOUND-
LAND,
Newfoundland, Canada

UNIVERSITY OF NEW
HAMPSHIRE,
Dept. Political Science,
Durham,
New Hampshire 03824, USA

UNIVERSITY OF NORTH
CAROLINA,
Chapel Hill,
North Carolina, USA

UNIVERSITY OF NOTTINGHAM,
Nottingham NG7 2RD,
Nottinghamshire

UNIVERSITY OF OREGON,
Eugene,
Oregon 97403, USA

UNIVERSITY OF PITTSBURGH,
Pittsburgh,
Pennsylvania, USA

UNIVERSITY OF SOUTHERN
CALIFORNIA,
University Park,
Los Angeles,
California 90007, USA

UNIVERSITY OF STRATHCLYDE,
Pitt Street,
Glasgow C.2

UNIVERSITY OF SURREY,
Guildford,
Surrey

UNIVERSITY OF TENNESSEE,
Knoxville,
Tennessee, USA

UNIVERSITY OF TEXAS,
Austin BEB 203,
Austin,
Texas 78712, USA

UNIVERSITY OF TOLEDO,
Toledo,
Ohio 43606, USA

UNIVERSITY OF TORONTO,
Toronto, Canada

UNIVERSITY OF WALES,
University College,
51 Park Place,
Cardiff

UNIVERSITY OF WASHINGTON,
Seattle,
Washington 98105, USA

URBAN AFFAIRS QUARTERLY,
SAGE Publications,
275 South Beverly Drive,
Beverly Hills,
California 90212, USA

URBANDYNE,
5659 S. Woodlawn Avenue,
Chicago,
Illinois 60637, USA

URBAN SYSTEMS INC,
1033 Massachusetts Avenue,
Cambridge,
Massachusetts 02139, USA

URWICK ORR AND PARTNERS
LTD,
Baylis House,
Stoke Pogis Lane,
Slough,
Buckinghamshire

RICHARD J. URWIN INC,
Homewood,
Illinois, USA

VETERANS ADMINISTRATION,
810 Vermont Avenue North West,
Washington D.C. 20420, USA

VICTORY,
J. W. Spear & Sons Ltd,
Green Street,
Enfield,
Middlesex EN3 7SF

VISITING NURSE SERVICE, N.Y.,
New York, USA

VISUAL EDUCATION,
33 Queen Anne Street,
London W.C.1

WGBH EDUCATIONAL FOUNDATION,
125 Western Avenue,
Boston,
Massachusetts 02134, USA

JOHN WADDINGTON LTD,
Morris House,
Berkeley Square,
London W.1

WAR OFFICE,
London Road,
Stanmore,
Middlesex

WARGAMES RESEARCH GROUP,
75, Ardingly Drive,
Goring-by-Sea,
Sussex

FREDERICK WARNE LTD,
40 Bedford Square,
London WC1B 3HE

JOHN E. WASHBURN,
P.O. Box 702,
San Anselmo,
California 94960, USA

WASHINGTON UNIVERSITY,
Box 1133,
St. Louis,
Mississippi 63130, USA

WAYNE STATE UNIVERSITY,
Detroit,
Michigan 48202, USA

WELLESLEY SCHOOL SYSTEM,
Wellesley,
Massachusetts 02181, USA

WESSEX REGIONAL HOSPITAL
BOARD,
High Croft,
Romsey Road,
Winchester

WESTCHESTER BOARD OF
CO-OPERATIVE EDUCATIONAL
SERVICES,
Yorktown Heights,
New York 10598, USA

WESTERN BEHAVIORAL
SCIENCES INSTITUTE,
1121 Torrey Pine's Boulevard,
La Jolla,
California 92037, USA

WESTERN CENTER,
9400 Culver Boulevard,
Suite 206,
Culver City,
California 90230, USA

WESTERN PUBLISHING
COMPANY INC,
850 Third Avenue,
New York,
New York 10022, USA

WESTINGHOUSE INFORMATION
SYSTEMS LABORATORY,
2040 Ardmore Boulevard,
Pittsburgh,
Pennsylvania 15221, USA

WFF'N PROOF INC,
Box 71,
New Haven,
Connecticut, USA

ROY F. WESTON,
Lewis Lane,
West Chester,
Pennsylvania 19380, USA

WHITMAN PUBLISHING CO,
560 West Lake Street,
Chicago,
Illinois 60606, USA

WHITWORTH COLLEGE,
Spokane,
Washington 99218, USA

J. WILEY & SONS LTD,
Glen House,
Baffins Lane,
Chichester,
Sussex

WILTSHIRE TRAINING AGENCY,
County Hall,
Trowbridge,
Wiltshire

WORKING TOGETHER CAMPAIGN,
Room G4, Portland House,
Stag Place,
London SW1E 5BJ

WORLD AFFAIRS COUNCIL OF
PHILADELPHIA,
Third Floor Gallery,
John Wanamaker Store,
13th and Market Streets,
Philadelphia,
Pennsylvania 19107, USA

WORLD LAW FUND,
11 West 42nd Street,
New York,
New York 10036, USA

XL,
16 Fairview Avenue,
Levershulme,
Manchester 20

YOUNG MENS CHRISTIAN
ASSOCIATION,
51, Victoria Street,
St. Albans,
Hertfordshire

YOUTH SERVICE INFORMATION
CENTRE,
37 Belvoir Street,
Leicester LE1 6SL,
Leicestershire

Incomplete addresses

We have been unable to trace the full addresses of the following manufacturers.

Benassi Enterprises
G. Bradford.   Canada
A. Calhamer.   USA
Championship Games
CUES Managers Society
S. Curtis
Electronic Data Controls
Ferguson
Greenfields High School
Imprimerie Centrale.   Sweden
Instructional Systems Inc.   USA
Instructional Developments Corporation.   USA

J. K. James.   UK
Langofun International
D. Lebling
R. Loomis
E. S. Lowe.   USA
Mills Center.   USA
Rolling Front Game Company.   USA
S.B. Modules Ltd.   UK
Solidarity House Inc.   USA
Southern Simulation Service.   USA
Sports Link Ltd.   UK
Teacher Research.   USA

5.5   Appendix

The register of games and simulations listed and indexed above is clearly not a completely exhaustive compilation. Only those games have been included on which sufficient information is available to allow them to be accurately cross-referenced. However, since it may be of value to the user of the Handbook to know the titles of games on which our records are incomplete a further alphabetical list is given below. It is hoped that readers who can supply additional details on any of these games will contact the author so that more comprehensive information can be provided in the next edition.

ABC Management Exercise Game
Accounting for Decisions
Accounting Information and Business
   Decisions
Activities Allocation Model
Acute Abdomen.   W. Saxbe.   USA
Adman
Advul

African Crisis Game
The 'A' Game
Agri-Business
Agriculture
Air Strategy Game
Airline Operating Game
Airline Sales Game
Aldous

Allegiance. USA. Western Behavioural Sciences Institute
Amstam Business Game
The Andlinger Game
The Antioch Game
Appalachian Airlines Simulation Game
Aquisition. 1972. UK. Ferguson
Around the World
Artificial Society
Assembly Line
Association. USA. Western Publishing Corp
Atlantis
Authors
Autospan
Bank Management Game. USA
Baselous. USA
Battlecry. USA
Battleship. USA
Big (Business Investment Game). USA
The Big Town. UK
Blitz. 1967
Board Game. USA
Boston College Decision Making Exercise. USA
Bristol. USA
Broadside. USA
Budgetary Politics and Presidential Decision Making. USA
Buloga (Business Logistics Decision Simulation). USA
Burroughs Economic Simulator. USA
Burroughs Supermarket Battle Maneuvers. USA
A Business Exercise for Section Supervisors and Division Managers. USA
Business Game III, Executive Decision Making. USA
Business in Action. USA
Business Logistics Facility Location Simulation. USA
Business Management Laboratory. USA
Business Management Simulation. USA
Business Simulation Game. USA
Business Simulator. USA
Business Strategy Simulation. USA
Capital Budgeting. USA
Casdon Soccer. UK
Cattle Ship. USA
Chemandments I. USA
Chemistry Lab Simulation. USA
Children's Hour. USA
Chutes and Ladders. USA

The City. USA
Classroom Experiences. USA
Collective Negotiations Simulation. USA
College Activities Game. USA
Common Market. USA
Compagnie Francaise D'Organisation Game. USA
Compass. USA. Instructional Systems Inc
Compatability. USA
Completed Staff Work. USA
Computer Marketing Game. USA
A Computer Orientated Game Simulating the Combined Production Scheduling-Inventory Control Problem. USA
Conformity Status Characteristics. USA
The Conopoly Industry. USA
Conquest. USA. Instructional Systems Inc
Control. USA. Instructional Systems Inc
Constitution Ratifying Convention. USA
Contack. USA
Contrast American Simulation. USA
Control Function. USA
Cooperation Game. USA
Counsellor Game. USA
Court Room
Cross-Cultural Communications Simulation. USA
Daedalus
Das Unospiel. Germany
Decimal Shopping. UK
Decision Exercise. USA
Decision-Making by Congressional Committee. USA
Decision Making Simulation. USA
Decisions of Inquiry. USA
Deep. USA
Defence Management System (DMS). USA
Destiny. USA. Instructional Systems Inc
Det-Detect Key Words. USA
Dial 'N' Spell. USA
Discovering How to Learn. USA
Dismissal of General Douglas MacArthur. USA
District Simulation Management Game. USA
Drug Wholesale Management. USA
Dynamic Executive Simulation. USA
Dynamo Aerospace Simulation Workshop. USA

East-West Conflict
Easy Math. USA
Easy Money. USA
Econometric Model of US. USA
Econometric Model of US Economy. USA
The Economics of Peace and War. USA
Educational and Career Planning
  Simulations. USA
Educational Concentration. USA
Educational Password Game. USA
Educational Planning Game. USA
Educational Problems Simulation. USA
Eedmas (Engineering Economics
  Decision-Making Simulation). USA
18th/19th Century Mercantilism
Electron '68. 1968. USA
Elf Concepts. USA
Emotional Tic-Tac-Toe. USA
English Civil War Games. USA
Environment Simulation. USA
The Establishment. USA
Executive Decision-Making Through
  Simulation. USA
Executive Decision—Model 1. USA
Executive Decision Simulation. USA
Exercise in Business Management. USA
Exercise in Money and Banking. USA
Expectation of Competance. USA
Familiar Things. USA
Family Decision Making Experience. USA
Farm Organization and Investment Game.
  USA
Federal Reserve Bank Simulations. USA
Finance. USA
Financial Management Game. USA
Firm Behavior Patterns. USA
Fix-It. USA. Science Research
  Associates, Abt Associates Inc
Focus: A Life Agency Management Game.
  USA
Foreign Policy Simulation. USA
Forest Fire Simulation. USA
Fort Simulation USA. USA
Foundation. USA
4 Cyte. USA
Fours. UK
Friedland. USA
Fuseg. USA
Future. USA
Game 21. USA
The Game of Conversational 'Leading'. USA
Game of Self Defense. USA

Game of the States. USA
The Gaming Company. USA
General Airline Game. USA
General Business Management Simulation.
  USA
Geography Lotto. USA
The Georgia-Pacific Management Game.
  USA
Go-Bang
Goal. UK
Goat. USA
Go to the Head of the Class. USA
Grand Tactical Game. A. L. Totten. USA
Great Circle I. USA
Great Circle II. USA
Guerilla Warfare Model. USA
Guerre D'Outrance
Gusher. USA
Harbus 2 Model. USA
Heads Up. USA
Hexapawn—Learning Program. USA
Hey Boss (Boss Man, Big Boss). Germany
High School. USA. Johns Hopkins
  University
Hit the Beach. USA
Homunculus. USA
Hostility. USA
A House in Ancient Greece. USA
Icarus
Ideom. USA
IFMA. USA
Illini Egg-Handler Simulation. USA
India. USA
Industrial Sales. USA
The Innovation Game. USA
Innovations for Real People. USA
Interaction. I. L. Preston. 1966. USA.
  Simulation Inc
Internal Psychological Conflict. USA
International Relations and Labour
  Management Games. USA
Interstate Commerce. USA
Introduction to Karl Marx. USA
Inventory Decision Game. USA
Inventrol. USA
Invitation to Inquiry. USA
'ISMS'. USA. Instructional Systems Inc
It's in the Bag. 1972. UK
The Japanese Family. USA
Jotto. USA
Judicial Decision Process Simulation. USA
Juezo de la Empresa. USA

Kalah. USA
Kansas/Nebraska Act. USA
Keep Them Off the Streets. T. K. Shutt. UK
Know Your Country. USA
Konsai Business Administration Association's Process Game. USA
Kroger Supermarket Decision Simulation. USA
Krypto. USA
Land Locked Nations Game. USA
Latin American Revolution
L-Check. 1972. UK. Fergusons Ltd
Legislature Game. USA
Lensman. P. Pritchard. USA
Life Chances and the Black Ghetto American. USA
A Life Insurance Company Management Game. USA
Lifetime Income Distribution Model. USA
Literary Analysis Simulation. USA
Local Politics Game. USA
Location. USA. Instructional Systems Inc
Look and Learn Lotto. USA
Maintenance Management Game. USA
A Management Decision Game. USA
Management Decision Game–Small Business. USA
Management Development. USA
A Management Game for the Petroleum Industry. USA
Management Game for the Natural Gas Transmission Industry. USA
Management Simulation Game. USA
A Managerial Game for an Insurance Company. USA
Manpower Simulator for High Talent Personnel. USA
Manufacturing Executive Game. USA
Mark XIV Executive Decision Simulation. USA
Market Negotiation. USA
Marketing Decisions Simulation. USA
Marketing Game. USA
Marketing in Action–A Decision Game. USA
Marketing Management Simulation. USA
Marketing Strategy Simulation. USA
Marketplace. USA
The Marman Management Game. USA
Match-Things. USA

Material Management Game. USA
Materials Inventory. USA
McKesson Wholesale Liquor Management. USA
Medium. USA
Meet the Presidents. USA
Merels.
Messages. USA
Micro–Analytic Model of a Socio-Economic System. USA
Micro Society. USA
Minor Tactical Game. A. L. Totten. USA
Minos. UK
Mississippi Plan. USA
Misslogs. USA
Miss World Game. 1972. UK
Mr. Freud. USA. Instructional Systems Inc
Model for Evaluating Transportation Systems. USA
Model for Instructional Systems Design. USA
Model of Investment. USA
Moon Explorer Problem. USA
MSU Investment Game. USA
MSU Management Game. USA
National Level Civil Defense. USA
National Policy Formation. USA
Naval Sea and Air Engagement. USA
Net Worth. USA
New Federalism. USA. Instructional Systems Inc
Numble. USA
Objectives I. USA. Instructional Systems Inc
Objectives II. USA. Instructional Systems Inc
Ohio River Basin Model. USA
O.I.D. Management Simulation Exercise. USA
The Oklahoma Executive Management Game. USA
Oklahoma Farm Management Decision Game II. USA
Opening the Deck. USA
Operation Federal Reserve. USA
Operation Interlock. USA
Operation Step-Up Simulations. USA
O.Q.O. USA. Instructional Systems Inc
Oregon Farm Management Simulation. USA
Organization-Oriented Game. USA

Our Working World. USA
Pacemaker Games Program. USA. Fearon
Parent–Satellite Nations. USA
Paris
Parm. USA
Perquacky. USA
Personnel Assignment. USA
Peso. USA
Petroleum Economy Simulation. USA
Phonetic Quizzes. USA
Physical Distribution Simulation. USA
The Pillsbury Company Management Game.
USA
Pirate and Traweller. USA
Pit. USA
Pitt Amstam Market Simulator. USA
Pittsburgh Urban Renewal Model. USA
Planning Models for Indian Development.
USA
Planning Simulation Exercise. USA
Plant Scheduling and Warehouse Distribution.
USA
Plato. USA. Instructional Systems Inc
Play 'N' Talk. USA
The Plot. UK
Police Training. USA
Polismog. USA. K. Bridbord
Political Convention
Portfolio Management Game. USA
Practice of Supervision. USA
Pre-College Training Simulations. USA
Presidential Election. USA
Presidential Election Campaigning. USA
Production Line. USA
Production Management Game. USA
Production Management Simulation. USA
Production-Manpower Decision Game. USA
Production Manpower Game. USA
Production Simulator. USA
Project Sobig. USA
Properties I. H. A. Springle. 1969.
USA. Science Research Associates
Properties II. H. A. Springle. 1969.
USA. Science Research Associates
Property and Liability Insurance Game. USA
Prospectville. USA
The Public Sector. USA
Pyramid. USA
Quads. A. Spear. USA
Questioneze. USA
Radix. USA
Reading Readiness. USA

Region. USA
Relationships. USA
Reporting. USA
Research and Development Effectiveness.
USA
Research and Development Game. USA
Retailing Department. USA
ROHR Business Game. USA
Roll A Puzzle. USA
Rolling Doughnut Metal Truck Company.
USA
Rotascript. Holland. Jumbo
Sales Management Simulation Exercise.
USA
San Francisco Community Renewal Project
Model. USA
Scam. 1972. USA
Scheduling Game. USA
School System Model. USA
Scriptogram. UK
Section Supervisors Game. USA
Self Perception. USA
Sentence. USA
Sesame. USA
Shake-A-Word. USA
Shelter Management Contingency Game.
USA
Simscript. USA. Southern Simulation
Service
Simuland. USA
Simulating Didactic Teaching. USA
Simulation Exercise for the Training of
Educational Evaluators. USA
Simulation for the Training of R & D
Project Managers. USA
Simulation of the Diesel Engine Combustion
Process. USA
Simulation of International Relations. USA
Simulation of the Venezuelan Economy.
USA
Simulett. USA
Simuload. USA
A Slimming Exercise. UK
The Small Town Community. UK
Solar Conquest. USA
SPARC (Space Planning Against Ranged
Contingencies). USA
Stanford University Simulation. USA
Stanford University
Starting-Things in Simulation. USA
States and Cities. USA
Stock Exchange Game–Sobig. USA

Striker. UK
The Sumer Game. USA
The Sun Game. M. S. Gordon. USA.
  Abt Associates Inc
Super A. USA
Supermarket Battle Maneuvers. USA
Supermarket Decision Simulation. USA
Surgical Schedulers' Management Game.
  R. Hoffman. USA
Survival. USA
Survival Game. USA
Systems Development Planning Structure.
  USA
A Tactical and Negotiations Game. USA
Taking A Walk. USA
The Task Manufacturing Corp. USA
Teacher Time Allocation Simulation. USA
Teaching Neuroanatomy. USA
Technological Innovation. USA
Television Game. USA
Tennis Anyone? USA. Franklin
  Merchandising
Ten Piccaninnies. UK
Tense-Things. USA
Think-A-Dot
A Three Dimensional Gaming Exercise for
  Dynamic Salesman-Manager Training.
  USA
Time of Your Life. USA
Tinker Toy-Objects Game. USA
Tool Game. USA
Top Brass. USA
Top Management Decision Simulation. USA
Top Management Game. USA

Top Team. UK
Trading Post. USA
Trafalgar. R. Cormier
Transcheck. 1972. UK. Ferguson
Traffic. USA
Transportation Management Simulation.
  USA
Trienline. 1972. UK. Fergusons
Two Person Non-Verbal Bargaining. USA
Tycoon
Ultima
Uniflo. USA
United Nations
United Nations Model. USA
Urbcom. USA
Urbos. USA
Using Financial Data in Business Decisions.
  USA
Versailles Treaty. USA
Virgin Islands Game. USA
Western Expansion. USA
The Wharton Executive Game. USA
Wholesale Building Material Simulation.
  USA
Why Modern Mathematics. USA
Wildcat. USA
Word Game. USA
Wordsmith Ltd. USA
World Politics Simulation. USA
World Soccer. T. Kremer. 1971. Hungary
The Village Community. UK
Zeno. USA. Instructional Systems Inc
Zoo Twins. USA